CELESTIAL PRAISE FOR

TIME FOR THE STARS

AND

ALAN LIGHTMAN

"A physicist with a decidedly poetic bent."

—***New York Times Book Review***

"Alan Lightman is a working theoretical astrophysicist with a flair for the informal and easy-to-read prose over an unusual range of stances and forms."

— **Philip Morrison, *Scientific American***

"Alan Lightman reminds me of the awesome Michael Crichton....He writes with a divine simplicity...discusses the basic scientific views on astronomy, offers an up-to-the-moment understanding of our solar system and planets....There is true brilliance at work here."

—***Coast Book Review Service***

more...

"Lightman is not a first-rate physicist who happens to be a good writer. He is a wonderful writer who happens to be a first-rate physicist."

—***Minneapolis Star Tribune***

"That it succeeds should be no surprise, given the credentials of the author...astronomy is as exciting an intellectual world as one can find. Lightman conveys that excitement...gives even science tyros a window on the universe."

—***Los Angeles Daily News***

"Lightman writes easily and gracefully, and this book will give any interested lay reader a clear guide to what astronomers of all kinds will be working on this decade."

—***Astronomy***

"Lightman's tone is that of an amiable observatory tour guide, one who tells less than he knows because he wants visitors to absorb the grandeur first."

—***Publishers Weekly***

"Lightman has an elegant, precise prose style that is well mannered, fluid, powerful, and gripping."

—***Memphis Commercial Appeal***

"A well-written history of astronomy, its strength lies in giving hints as to what the immediate future may hold."

—***Booklist***

"Important....Reporting the latest in astronomical discovery and research, Lightman surveys his field with pardonable pride, sharing its triumphs and sense of purpose, and predicting that surprises, as always, will be the most important discoveries of the future."

—***Anniston Star***

OTHER BOOKS BY ALAN LIGHTMAN

Problem Book in Relativity and Gravitation (1975)

Radiative Processes in Astrophysics (1979)

Time Travel and Papa Joe's Pipe. Essays on the Human Side of Science (1984)

A Modern Day Yankee in a Connecticut Court, and Other Essays on Science (1986)

Origins: the Lives and Worlds of Modern Cosmologists (1990)

Ancient Light: Our Changing View of the Universe (1991)

Great Ideas in Physics (1991)

TIME FOR THE STARS

ASTRONOMY IN THE 1990s

ALAN LIGHTMAN

A Time Warner Company

Warner Books Edition

This Warner Books edition is published by arrangement with Penguin Books USA Inc., 375 Hudson Street, New York, NY 10014

Warner Books, Inc., 1271 Avenue of the Americas, New York, NY 10020

A Time Warner Company

Printed in the United States of America
First Warner Books Printing: February 1994
10 9 8 7 6 5 4 3 2 1

Library of Congress Cataloging-in-Publication Data
Lightman, Alan P.
Time for the Stars: astronomy in the 1990s/Alan Lightman.
p. cm.
Originally published: New York, N.Y., U.S.A. : Viking, 1992.
Includes index.
ISBN 0-446-67024-3
1. Astronomy--History. 2. Astronomy and state--United States
I. Title.
QB15.L54 1993
520' .973--dc20 93-27402
CIP

Cover Design by Julia Kushnirsky

Foreword

Alan Lightman served as chair of the science panel for the National Research Council Astronomy and Astrophysics Survey for the 1990s, and was the principal author of a popular chapter on science opportunities in the council's report. As chair of the panel, he was able to observe how the American astronomy community posed the most crucial questions for the coming decade and arrived at the highest priority instrumental initiatives to help resolve these problems.

Lightman listened carefully to the debates about how best to find planets around other stars and the most promising way to discover the true nature of dark matter. He understood when observations best reveal how galaxies form and how quasars shine.

All this and much more is documented in this short book. In *Time for the Stars,* the reader will be confronted with puzzles about people and planets, about stars and galaxies, and about the beginnings and the ends of stellar systems and of the universe.

Alan Lightman is an exceptional research scientist who writes with simplicity, clarity, and verve. Above all else, he tells a good story. This book is what I want my children to know about astronomy and about what I do.

—John N. Bahcall,
Professor of Astrophysics,
Institute for Advanced Study, Princeton
Chair, Astronomy & Astrophysics Decade Survey Committee for the 1990s
President, American Astronomical Society

Preface

The title of this book is taken from a Robert Heinlein novel about a foundation dedicated to very long-term scientific projects. The foundation prided itself on funding only projects whose prospective benefits lay at least two centuries away. To the dismay and embarrassment of the directors, however, the foundation's most preposterous enterprises quickly began piling up large profits.

Like Heinlein's fictitious foundation, the U.S. National Academy of Sciences also tries to plan ahead, although not as far as two centuries. A private organization of leading scientists, the National Academy of Sciences was created in 1863 to provide expert advice to the government on matters of science and technology. At the beginning of each decade since 1960, the Academy has appointed a group of scientists, called the Astronomy & Astrophysics Survey Committee, to assess the state of American astronomy and to recommend new initiatives for the coming decade. The most recent Committee, charged with a plan for the 1990s, was chaired by John

Bahcall of the Institute for Advanced Study in Princeton. In the past, about two-thirds of the recommended projects have been funded by the U.S. Congress and eventually put into operation, although for the bigger projects this process can take as long as twenty years.

For the 1990 Astronomy and Astrophysics Survey Committee and its report, I was asked to write a broad summary of the basic scientific issues and recommendations. This book is based on that summary. Since astronomy has helped shape our worldview throughout history, I have included some historical and cultural background to place astronomy in its proper human perspective. I have also added biographical sketches and personal statements from some contemporary scientists, to give a feeling for the astronomers themselves.

Time for the Stars limits itself to science. Clearly, additional considerations are critical in putting recommended programs into operation, such as price tags, budgets, time frames, national priorities, and institutional control. I will briefly mention some of these other issues here. Roughly speaking, scientific projects in astronomy can be broken down into three categories: "small" projects, costing less than $100 million, such as conventional ground-based telescopes; "medium" projects, costing in the range of $100–$250 million; and "large" projects costing over $250 million. The Earth-orbiting observatories, the first two of which are the Hubble Space Telescope and the Gamma Ray Observatory, cost in the range of $1–$2 billion. (The proposed Advanced X-ray Astrophysics Facility and the Space Infrared Telescope Facility also fit into this category.) Astronomers, and indeed all scientists these days, debate the relative virtues of "big science" versus "little science," that is, doing science with a small number of relatively expensive instruments of unique capabilities versus doing science with a large number of cheap but less capable instruments. Astronomers also debate the merits of international collaborations and the relative roles of national versus private facilities. For space-based missions, astronomers debate the merits of various kinds of launch vehicles.

Some of the scientific projects now being recommended for the 1990s and beyond will not be put into action. And certainly the time frames may change. Therefore, I have not mentioned in the text definite dates and budgets of proposed instruments. However, a table at the end of the book lists approximate dates of operation for some of the proposed projects, as well as information on some of the astronomical instruments of the recent past.

The astronomical community, and the world, were shaken by the failure of the focusing ability of the Hubble Space Telescope, launched in April 1990 after three decades of planning. An investigation of the failure has shown that the problem did not arise from the intrinsic design or specifications of the telescope, but rather from an inaccurate shaping of the telescope's mirror and inadequate testing of that shape. Thus the disaster, as painful as it has been, does not herald a fundamental flaw in the capability of advanced astronomical equipment. The long-term goals and aspirations of astronomy have not changed. We must have time for the stars.

In preparing this book, I had a great deal of help from my scientific colleagues. I am particularly grateful to John Bahcall, Sallie Baliunas, Charles Beichman, Roger Blandford, Alastair Cameron, Marc Davis, James Elliot, George Field, Fred Gillett, Paul Horowitz, Garth Illingworth, Kenneth Kellerman, Bruce Margon, Brian Marsden, Christopher McKee, David Morrison, Philip Myers, Stephen Myers, Robert Noyes, Jeremiah Ostriker, Stephen Ridgway, Robert Rosner, Vera Rubin, Paul Schechter, David Schramm, Irwin Shapiro, Alar Toomre, Michael Turner, Steven Willner, Sidney Wolff, and Edward Wright. For his fine editorial suggestions, I thank my editor Michael Millman. Of course, I must take responsibility for any factual errors remaining in the book. Finally, the relative emphasis of various proposed projects in this book does not necessarily reflect the priorities of the National Academy of Sciences.

Contents

INTRODUCTION xv

OUR SOLAR SYSTEM AND
THE SEARCH FOR OTHER PLANETS 1

The Formation and Evolution of Our Solar System 1

The Search for Other Planets 13

The Search for Extraterrestrial Intelligence 21

THE LIFE HISTORY OF STARS 23

The Sun 23

The Formation of Stars 29

The Life and Death of Stars 32

THE LIFE HISTORY OF GALAXIES 49

The Discovery of Galaxies 49

The Evolution of Galaxies 56

The Power Source of Quasars and Active Galaxies 62

The Birth of Galaxies 70

THE LIFE HISTORY OF THE UNIVERSE 77

The Big Bang Model 77

The Large-Scale Structure of the Universe 83

Dark Matter 89

The Origin of the Universe 95

The End of the Universe 102

Table of Recent and Proposed Astronomical Instruments 109

Illustration Credits 113

Index 115

Introduction

According to the ancient Babylonian story of creation, *The Enuma Elish,* we can trace ourselves back to our parents and grandparents and great-grandparents and so on until we get back to Anu, the sky god, and Nudimmut, the earth god, who were formed from the slow seep of silt at the liquid horizon. Before that there was watery chaos.

Modern astronomers believe that the atoms of our bodies came from beyond the horizon, formed by the nuclear reactions inside stars and then blown into space to mold planets, soil, and organic molecules. This new view of our human beginnings includes with it a life history of stars and galaxies and even the whole universe. While astronomers of the past were concerned more with charting the stars in a permanent cosmos, astronomers today study evolution and change.

Throughout history, astronomy has shaped our worldview. The seasons and motions of the heavenly bodies were the earliest examples of regularity in nature. Astronomy thus became the first science. The astrolabe of the ancient Greeks served as a timekeeping device, later yielding to other clocks that acknowledged a reality of time outside of human perception. The "sight and grid" used by Albrecht Dürer and other artists first studying perspective came from the wooden astronomical sights used to locate stars. In the sixteenth century, the sun-centered planetary system of Nicholas Copernicus dislodged Earth from its privileged position and challenged the view that the cosmos was made for human beings. Isaac Newton's law of gravitation, gleaned from the orbits of planets, was the first modern theory of nature and the inspiration for many later theories. Edwin Hubble's discovery in 1929 of the expansion of the universe destroyed the Aristotelian belief in an unchanging cosmos.

In a celestial echo of Darwin's work in the last century, astronomical findings in this century have raised questions about evolutionary processes in space. We have found active volcanoes on Venus. We have recognized the raw material of the solar system, frozen in comets for billions of years. We have witnessed stars in the act of formation, still wrapped in the gas and the dust out of which they condensed. We have seen other stars exploding, having first spent their nuclear fuel and collapsed under their own weight. And in the stellar debris, we have found oxygen and carbon and other such elements essential to life. We have discovered huge streams of matter propelled from the centers of galaxies at nearly the speed of light. We have observed variations in the color and luminosity of galaxies at different stages of their evolution. We have learned that galaxies are not scattered evenly in space, as once believed, but bunch together in filaments and sheets and other large groupings whose origins have not been explained. Finally, we have gathered evidence that the entire mass of the universe began in a state of fantastic compression, some 10 billion years ago. How did the universe come into being in the first place? What determined its properties? Will it keep expanding forever or instead recollapse? A century ago, such questions were considered to lie outside the domain of science. Today, they are at its core. We now recognize that all things in the cosmos undergo change.

Many of the new findings have been driven by advances in technology. In the 1930s, new communication devices led to the reception of radio waves from space. For thousands of years before, visible light had been humankind's only way of seeing the world. Since the 1940s, a series of rockets and satellites have recorded infrared radiation, ultraviolet radiation, and X-rays emitted from space. Such radiations, like radio waves invisible to the human eye, have revealed completely new features of many astronomical objects and announced some objects not before known. Electronic light detectors have replaced photographic plates, recording images of astronomical objects in one hundredth the exposure time and allowing those images to be stored and manipulated by computers. Other high-speed computers have revolutionized theoretical astronomy by permitting the simulation of millions of interacting particles, each representing an electron or a star or a galaxy. Large arrays of radio telescopes, electronically linked to combine the data from different antennas, can work together as if they were one giant eye.

When I was in high school, in the mid-1960s, astronomers sat in a cage or stood on a platform and guided the telescope *by eye*. Now guiding is done by a machine, electronically. In the 1960s, astronomers took their data in photographs, made by a camera at the back end of a telescope. Today, you can't find a photographic plate near a large telescope. The faint light of stars and galaxies is recorded by advanced photoelectric cells, called charge-coupled devices, and stored in a computer. In the 1960s, most of the astronomers I heard about worked with visible light—light that the human eye could see. I had the notion that astronomers on "observing runs" would pack up several days' worth of sandwiches and good books for the cloudy nights, travel to the top of a mountain somewhere, and sit at the eyepiece of a telescope, enjoying the starry spectacle firsthand. Today, there are large groups of "X-ray astronomers" and "infrared astronomers," and observing runs are often done by remote control. Recently, a colleague of mine, working with data from the Einstein X-ray Observatory (a satellite) completed a "hands-on" investigation of quasars. When I asked him what it was like, he said that he had passed the time in front of a video screen, pushing keys and pondering various digitized images of the quasar from data stored on a magnetic tape. The information had previously been manipulated by two other computers, after it had been beamed down to

Earth from the Observatory. It was irrelevant that the Observatory itself, which alone "saw" the quasar, in X-ray light, had been defunct for a decade. Digitized data keep well.

In the 1990s, astronomical exploration will also take advantage of novel technology and instruments now being planned. However, in equipping ourselves for the future, flexibility must accompany precision. We will theorize and forecast as well as we can, but if the past is a guide, some of the next decade's discoveries will catch us off guard. In astronomy, the frontiers surround us.

TIME FOR THE STARS

Our Solar System and the Search for Other Planets

The Formation and Evolution of Our Solar System

"Planet" means wanderer in Greek. So named because they seemed to wander from night to night through the fixed stars, five planets were known to the ancients: Mercury, Venus, Mars, Jupiter, and Saturn. For Aristotle, the planets were divine and ageless bodies; for Democritus, they were accidental clusterings of atoms. The kinship of the planets to Earth became apparent in 1610, when Galileo used the first telescope to spot moons around Jupiter. Uranus was discovered in 1781 by an unexpected sighting with a telescope; Neptune was found in 1846, after it had been predicted by theoretical calculations based on the orbital motions of Uranus; and Pluto was sighted in 1930.

To be called a planet, an object must have a mass less than about one-tenth that of our sun. A body with greater mass would be sufficiently hot to kindle its nuclear fuel and shine on its own; it would be a star. Within this restriction in mass, planets come in a range of sizes and chemical makeups. Jupiter, the largest planet in our solar system, has a mass 320 times that of the Earth and is almost all hydrogen and helium. Pluto, the smallest, has a mass 400 times smaller than the Earth's. The Earth itself has a mass about 6 trillion trillion kilograms (about 300,000 times smaller than the sun's mass), a diameter of about 8,000 miles, and a density of five grams per cubic centimeter (five times that of water). Earth's distance from the sun is about 100 million miles. Pluto, on average the most distant planet, is about forty times farther away.

The planets of our solar system divide into two groups: the "terrestrial" planets, Mercury, Venus, Earth, and Mars; and the "Jovian" planets, Jupiter, Saturn, Uranus, and Neptune. The terrestrial planets, which are those closest to the sun, are smaller, denser, and composed of rocky and metallic material. The Jovian planets are larger, less dense, and composed mostly of hydrogen, like the sun. Tiny Pluto is sometimes classified with the terrestrial planets.

In the middle of the eighteenth century, the German philosopher Immanuel Kant proposed that our system of planets and sun condensed out of a great rotating cloud of gas and dust. This proposal, called the nebular hypothesis, is still favored today. The primitive gaseous cloud would have slowly contracted, under the inward pull of its own gravity. Its central, densest regions formed the sun. Its outer regions would have collapsed downward along the axis of rotation, because of gravitational forces, but could not have fallen directly toward the primeval sun, because of centrifugal forces pushing outward. Caught between these two forces, the material would have formed a flattened disk, called a protoplanetary disk, in orbit about the sun. The part of the protoplanetary disk closest to the sun would have been hotter. Therefore, some of the volatile gases, such as hydrogen, would have been unable to condense into "ices" in the inner regions but could have done so farther out, while rocks and metals could have condensed even in the high heat near the sun. Such considerations partly explain the difference in composition between the inner and outer planets. For the last three decades, theoretical astronomers have used computer simulations to study how the gas and particles in our protoplanetary disk might have coalesced into planets and moons.

The nebular hypothesis was given observational support in 1983, when an orbiting telescope in space, the Infrared Astronomical Satellite, found the first evidence for disks of particles orbiting stars. To date, such disks of particles, which may be the debris left over from the formation of planets, have been found around as many as a quarter of all nearby stars. Also in the 1980s, coordinated telescopes sensitive to cosmic radio waves found protoplanetary disks orbiting around young stars. Many questions remain. What is the nature of the orbiting particles? What role do these particles play in planet formation? Exactly how do protoplanetary disks form planets, and how common is the phenomenon?

Until the twentieth century, a mainstay of Western thinking, inherited in part from Aristotle, was the notion of a static cosmos. For example, in the nineteenth century the Uniformitarian school of geology, led by British geologist Charles Lyell, argued that the Earth in the past was pretty much as it is today, unaging. However, evidence began to show otherwise. On his trip to Santiago in the Cape Verde Islands in the early 1830s, Charles Darwin noted that the deeper rocks were crystalline and volcanic, while the rocks near the surface were made of lime. Sediments of shale and limestone appeared to accumulate in time. Fossilized remains of animals and plants bore testimony of a world no longer seen. The continents have been shown to drift and separate over the ages, as first proposed by Alfred Wegener in the early twentieth century. And, beginning with the work of British physicist William Thomson in the mid-nineteenth century, we have come to realize that the sun, which heats the Earth, does not itself have an inexhaustible supply of heat. By the beginning of the twentieth century, we could no longer hold the view that the Earth has not evolved.

At the beginning of this century, chemists used the proportions of uranium and lead in rocks to date the Earth at a few billion years old. The modern value is 4.5 billion years. Four and a half billion years ago, the Earth condensed out of the gas and particles orbiting the sun. Four and a half billion years ago, in the primeval oceans and air of our planet, what molecules were present to make the amino acids and proteins and tissues of the first living creatures?

What was the origin of life? More generally, what was the primitive chemical composition of our solar system at the time of its birth?

We cannot answer these questions simply by digging underground. Planets and moons, heated by the sun and squeezed by their own weight, have melted and congealed and thus blurred the record of their birth. Comets, on the other hand, spend most of their time far from the sun and have little weight to support. It is to the comets, therefore, that we should look to discover the pristine material of our world.

The spacecraft missions to Halley's comet in the 1980s learned something of the primeval chemical composition of the solar system. In particular, astronomers discovered that the carbon in Halley exists in the form of large blobs of complex molecules, rather than in the far more simple form of methane and carbon monoxide gas, as previously believed. There is still much more to learn about the primitive forms of oxygen, hydrogen, carbon, sulfur, and nitrogen, especially those molecular forms of biological significance.

New astronomical missions and instruments proposed for the 1990s should advance our understanding of the structure and makeup of comets. The Comet Rendezvous Asteroid Flyby will be the first new spacecraft mission scheduled to explore comets at close range. On the ground will be the Millimeter Array (MMA), and in orbit about Earth will be the Space Infrared Telescope Facility (SIRTF). These instruments, and others, will be able to determine the size and makeup of comets as far away as Jupiter. The MMA will detect radio waves; SIRTF will detect infrared radiation. Too cold to emit visible light, the molecules in comets are still energetic enough to vibrate and rotate. Molecular vibrations produce infrared radiation; rotations produce radio waves. Both of these types of radiation, although too long in wavelength to be seen with the human eye, come in a range of "colors," and each kind of molecule imprints its unique identity in the particular spectrum of radio waves and infrared radiation it emits.

Indeed, a number of recent instruments in astronomy can detect radiation that is invisible to the human eye. Radiation we can see, called visible light, makes up just a small fraction of the "electromagnetic spectrum." Infrared radiation has longer wavelengths than visible light, and radio waves have still longer wave-

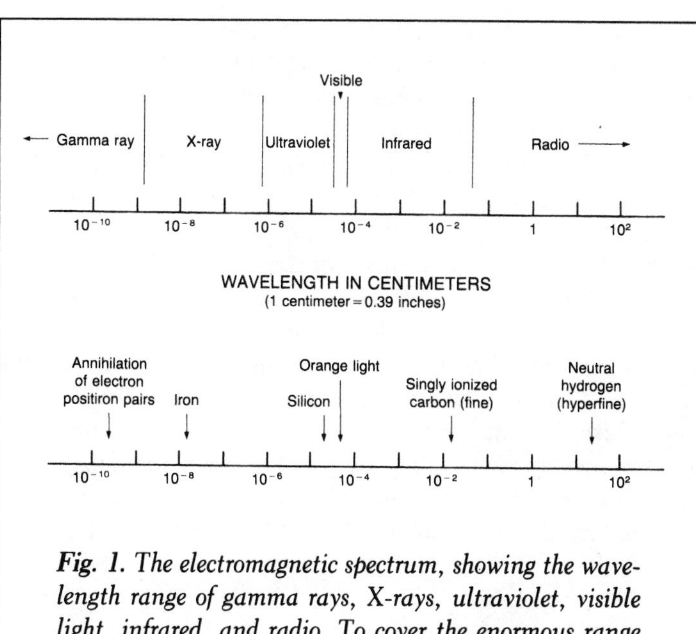

Fig. 1. *The electromagnetic spectrum, showing the wavelength range of gamma rays, X-rays, ultraviolet, visible light, infrared, and radio. To cover the enormous range of wavelengths, a scale is used in which each unit to the right increases by a multiple of ten rather than by a constant additive amount. Shown also are examples of specific atoms and molecules that produce electromagnetic radiation in each of the six wavelength ranges.*

lengths. On the other side of the spectrum, with wavelengths shorter than visible light, come ultraviolet radiation, X-rays, and gamma rays, in a succession of ever shorter wavelengths. Although these various radiations have different names, they are all similar forms of energy, differing only in their wavelengths. For the visible region of the electromagnetic spectrum, the different wavelengths correspond to different colors. Blue light has the shortest wavelength of visible light, and red the longest. Just as visible light can be separated into its component colors by a prism, other electromagnetic radiations, such as X-rays or infrared radiation, can be split into their component wavelengths. The amount of energy in each wavelength of incoming radiation is a fingerprint of the kind of atoms and molecules that produced that radiation. Unfortunately, only radio waves and visible light pass freely through the Earth's atmosphere, without absorption.

The wavelengths of electromagnetic radiation range from less than 0.000000001 centimeter (one centimeter equals about 0.39 inches) for gamma rays to greater than 0.03 centimeter for radio waves. To refer easily to such a vast range of numbers, from the very small to the very big, scientists use a numerical shorthand called scientific notation. For example, the number 0.000000001 can be denoted by 10^{-9}. The superscript "-9" means that there are nine zeros *before* the 1 (including the zero to the left of the decimal point). Likewise, the number 0.001 can be represented by 10^{-3}. Scientific notation may also be used to denote numbers larger than one by dropping the minus sign above the 10. For example, the number 10,000 can be represented by 10^4, where the 4 means that there are four zeroes *after* the 1.

Galactic and extragalactic astronomers must base their theories of the world entirely on the feeble light that arrives from distant space, but planetary astronomers can travel to their province. In the last three decades, the Mariner, Pioneer, Viking, and Voyager spacecraft have passed within a mere 10,000 miles of most of the planets of our solar system, relaying back startling information about new rings around Saturn, the soil of Mars, new moons around Uranus, the atmospheres of Jupiter and Saturn and Venus, a magnetic field around Uranus, and active volcanoes on a moon of Jupiter.

The findings from these missions have many implications, both for basic science and for understanding our own planet. For example, the study of atmospheres of planets in our solar system might possibly help us prevent man-made climate changes on Earth. The atmosphere of Venus is almost entirely carbon dioxide. This gas allows sunlight to come in, but prevents heat from escaping, thus acting as an enormous blanket around the planet. As a result, the surface of the planet heats up to a whopping 400 degrees centigrade. This blanketing phenomenon is called the "greenhouse effect." Venus may have once had a more Earth-like atmosphere (mostly oxygen and nitrogen), letting its heat out more easily and enjoying a much lower temperature. How did Venus evolve to its present situation? Information from spacecraft and other observations from the ground may help answer this question.

During the mid-to-late 1970s, spacecraft from the Viking mission landed on Mars and searched for life. Of all the planets in our solar system, Mars has conditions most similar to Earth's, and so

Fig. 2. Craters on Venus photographed by radar from the Magellan spacecraft. Craters shown range in size from twenty-three miles to thirty miles. Also seen are smaller domes of volcanic origin.

scientists believed that life might exist on Mars. Robots on the Martian surface scooped up Martian soil and tested it for living microorganisms. None were found.

In 1979, the Voyager spacecraft unexpectedly found erupting volcanoes on Io, one of Jupiter's moons. Indeed, Io appears to be the most volcanically active body in the solar system. The eruptions are so frequent and extensive that the entire surface of Io has apparently been covered over by sulfuric lava many times during the lifetime of the solar system. Careful study of the volcanoes has allowed estimates of the rate of heat flow from the interior of the moon. Where does the energy come from? Radioactivity within the interior of Io does not seem to be sufficient. Instead, astronomers hypothesize that the source of volcanic energy is tidal friction on Io from the gravitational pulls of Jupiter and its other moons. Further studies of the infrared and radio emission from Io should disclose the exact nature of the volcanic debris.

At the present time, two spacecraft, Magellan and Galileo, are on their way through the solar system. The Magellan mission arrived at Venus in August 1990. Using radar to penetrate the thick atmosphere and clouds, the spacecraft made a map of the planet. Scientists were astounded to find numerous craters, volcanoes with flowing lava, and regions with bright fracture lines running parallel to each other, about a mile apart. The causes of such strong geological activity are still unknown. The Galileo mission will arrive at Jupiter in 1995.

Planetary astronomers have by no means depended on space missions to explore the solar system. For example, in 1977, with a telescope aboard a specially designed aircraft, James Elliot and colleagues, then at Cornell University, discovered rings around Uranus. A year later, using a telescope on the ground, James Christy of the U.S. Naval Observatory discovered the moon of Pluto, named Charon after the mythological boatman who carried dead souls across the river of the underworld for judgment by Pluto. Careful observation of the orbit of Charon revealed for the first time the mass and size of Pluto: $\frac{1}{400}$ the Earth's mass and $\frac{1}{4}$ the Earth's diameter. Because of its low density, $\frac{1}{7}$ that of Earth, astronomers believe that Pluto is composed of solid methane, rather than the denser rocky material of Earth.

James Elliot was born on June 17, 1943, in Columbus, Ohio. After receiving an undergraduate degree in physics from the Massachusetts Institute of Technology in 1965 and a Ph.D. in astronomy from Harvard in 1972, Elliot did postdoctoral work at Cornell. In 1978, Elliot returned to M.I.T., where he is professor of planetary astronomy and director of the George R. Wallace Jr. Astrophysical Observatory. Elliot's main research interests involve probing the rings and atmospheres of the outer planets, which he does by carefully monitoring the changes in brightness of a star when a planet moves in front of it. Elliot and his colleagues discovered the rings of Uranus in 1977 and the atmosphere of Pluto in 1988.

Says Elliot, "Spacecraft visits have given us remarkably detailed information about most of the planets. Yet many critical questions about the solar system can be answered only through observations with telescopes on the ground or in Earth orbit. Do undetected bodies, perhaps even another planet, orbit our sun beyond Pluto? Do other stars have planets? If so, do these planets harbor life? Exactly how did the asteroids, comets, and ring systems form and evolve? What causes the weather and climate to change on the planets? The new telescopes and detectors of the 1990s will help us answer these questions."

For the future, the Stratospheric Observatory for Far-Infrared Astronomy (SOFIA), another proposed infrared telescope, should provide definitive information about Pluto's atmosphere, as well as the atmospheres and surfaces of other planets. SOFIA will be carried by an aircraft 45,000 feet above the Earth. SOFIA's relatively large light-gathering mirrors and other instruments will break apart incoming light into its telltale wavelengths with particularly high accuracy. As one of its missions, SOFIA will determine the various kinds of minerals on Mars and their locations, thus providing a geologic history of that planet.

Closer to home, how has the changing activity of the sun affected our own planet? On the basis of theoretical calculations, we believe that several billion years ago the sun was about 25 percent less luminous than today. Was the Earth's temperature lower as a consequence? Or was the lower solar luminosity compensated for by a terrestrial atmosphere that was more opaque and thus better able to trap the sun's warmth? We also have evidence that the newly formed sun spewed away its surface layers, which flowed outward from the sun in all directions like a great wind through the solar system. Did this powerful solar "wind" sweep out dust between the planets or change the primal atmosphere of Earth? How did it affect the climate? Little is known about these questions. In recent years scientists have proposed that Earth's climate could also have been altered through history by bombardment with comets throwing up huge clouds of dust and blocking the sun.

Viewed through the lens of change instead of constancy, old phenomena are seen in new ways. A good example is the Great Red Spot of Jupiter. First observed 300 years ago, the Great Red Spot is a reddish oval 20,000 miles in diameter, about a fifth of the planet's entire diameter. For centuries, astronomers thought the Great Red Spot was a painted fixture on the planet. In recent years, however, scientists have realized that the Spot floats in Jupiter's atmosphere and is constantly in motion, rotating counterclockwise once every six days. In 1978 closeup photographs from the Voyager spacecraft revealed much more. The Spot is part of a counter-rotating cyclone, a violent swirl of winds and fluids, churning round and round. In the midst of such turbulence, how could the Spot hold together for hundreds of years? The answer was provided by theorist Philip Marcus of the University of California at Berkeley and others and has involved a new field of science called nonlinear dynamics, or "chaos" theory. Although chaotic phenomena are still not well understood, the theory shows that islands of stability can miraculously exist amid chaos around them. The Great Red Spot is such a haven.

Philip Marcus was an undergraduate at the California Institute of Technology, where he got a strong background in theoretical physics, and then did his graduate work at Princeton. Marcus's Ph.D. thesis concerned the behavior of turbulent fluids, especially in the outer layers of stars. The theory of such fluids is extremely complex, and Marcus developed a simplified computer simulation of turbulence to help him understand what was happening. He also began comparing his computer results to other scientists' laboratory experiments with turbulent fluids, although such experiments involved conditions very different from those in stars. At the time Marcus was working on his Ph.D. thesis, in the mid- to late 1970s, two things happened. First, chaos theory, which had been quietly discussed by a small priesthood of scientists for a decade, was breaking out into the general ranks and becoming a hot topic. One of the central features of chaos theory is that realistic systems in nature are much more complex than the simplified, "linear" equations scientists had been using. For example, when "nonlinear" effects are included, small changes in the initial conditions of a physical system can cause big changes in the way it responds, as is often observed. Marcus put nonlinear equations into his computer programs. The second development was Voyager and the photos of Jupiter. Marcus retooled his computer program to study the Great Red Spot of Jupiter. Soon, using recent advances in computer graphics, he began displaying color movies of hypothetical planetary atmospheres, complete with spots and smudges that persisted amid the swirling gases around them. Still in its infancy, chaos theory has provided a unified understanding of many sciences, from physics to astronomy to biology.

Another long-standing and related astronomical question is

whether our entire solar system can hold together indefinitely, given the constantly changing gravitational pulls of the planets on each other. In the seventeenth century, Isaac Newton, whose theory of gravity allowed him to consider this problem, believed that "irregularities" in planetary orbits would mount and mount without occasional "reformations from God." A century later, the French mathematician and physicist Pierre-Simon Laplace employed an arsenal of mathematical calculations to claim with delight that the system of the planets could indeed govern itself in peaceful stability without any help from God. In the last decade, new methods of mathematical analysis and specially designed computers have been brought to bear on this difficult question. Although the answer is not yet known, Jack Wisdom of the Massachusetts Institute of Technology and others have found in their theoretical calculations that smaller bodies in the solar system can travel or tumble erratically for periods of time, then move in a stable and regular way, then return to chaotic behavior.

Gravity plays a crucial, if not so exotic, role in another puzzle tackled by theoretical astronomers: the rings around planets. The beautiful rings of Saturn have been admired since the early seventeenth century and consist of small particles in orbit about the planet. In the late 1970s, when astronomers discovered rings around Uranus as well, they found them to be quite narrow, and occurring only at certain distances from the planet. Why? In 1980 Peter Goldreich of the California Institute of Technology and Scott Tremaine of the University of Toronto proposed the existence of several previously unknown moons around Uranus. In theory, the new moons would act as "shepherds," gathering up the orbiting particles in some locations through gravitational forces. The Voyager spacecraft mission to Uranus subsequently found some of the hypothesized moons, with roughly the masses and positions predicted. (A previous confirmation of the theory had come unexpectedly, when the spacecraft first flew by Saturn and discovered a new ring, attended by a shepherding moon on each side of it.) For the 1990s, our theories of planetary rings will be challenged by the rings of Neptune. At present, not much data on these rings are available, and what we do have has not yet been fully explained.

Fig. 3. Rings of Saturn photographed by the Voyager spacecraft. Different shades indicate different material composition.

The Search for Other Planets

In the late sixteenth century, the Italian philosopher Giordano Bruno argued that space is filled with infinite numbers of planetary systems, inhabited by a multitude of living creatures. For this and other indiscretions, Bruno was burned at the stake. Yet the question remains: Are there other planetary systems in the universe?

In the 1980s, for the first time, some evidence was found for disks of material surrounding other stars. Unfortunately, such observations are difficult and the evidence is marginal. To date, we are not certain that *any* planet orbits a star other than our own sun. However, new instruments in the 1990s will almost certainly identify other planetary systems, if they exist.

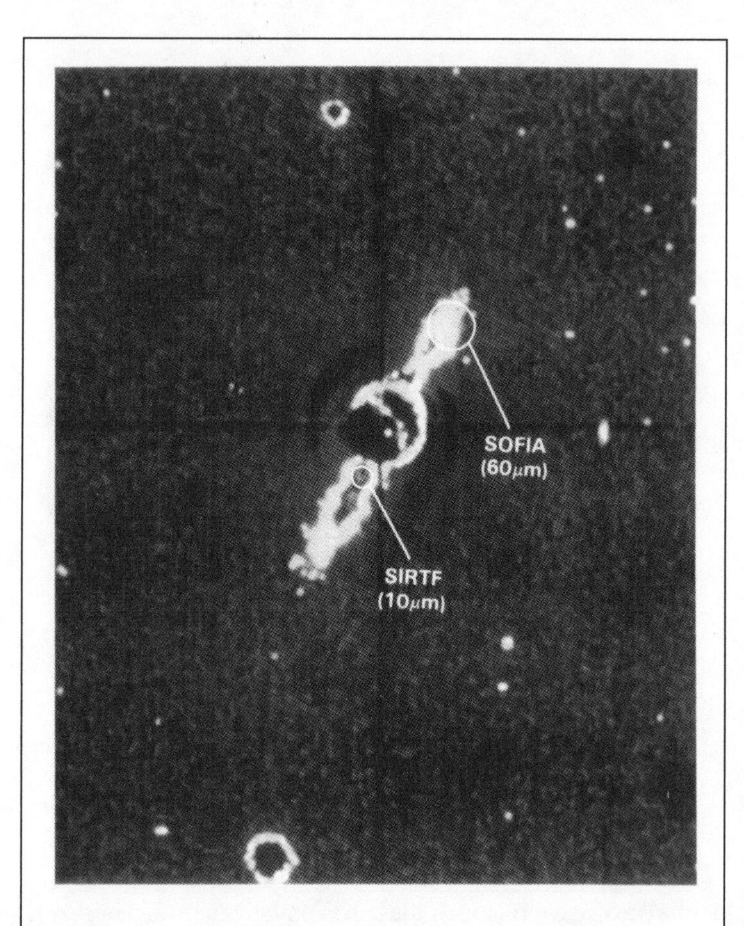

Fig. 4. A photograph made in visible light of the star Beta Pictoris, about sixty light-years from Earth. The disk around the central bulge is solid matter in orbit about the star, suggestive of a planetary system in formation. The disk emits strongly in the infrared. The two overlaid circles show the angular resolution, or "beam sizes," that will be achievable by the proposed SIRTF and SOFIA instruments in obtaining more detailed photos of the disk.

Direct detection of distant planets is not easy. The planets in our own solar system are easily seen by the light they reflect from the sun—during the night, when the sun is down and hidden from view. However, when viewing *other* solar systems from Earth, the central star is never hidden. Its light will dominate and obscure the light of any orbiting planets, just as a searchlight would overpower a firefly nearby. In addition, at large distances a central star and its orbiting planet will appear so close together as to be almost indistinguishable.

Fortunately, there are other ways to find planets. Planets Jupiter's size and larger emit their own radiation, with the required energy coming from their slow contraction and release of gravitational energy. This radiation emerges at infrared wavelengths. By contrast, most stars emit only a small fraction of their luminosity as infrared radiation. Thus at such wavelengths planets should be easier to distinguish from nearby stars.

The proposed Space Infrared Telescope Facility (SIRTF) could detect Jupiter in a solar system identical to our own if that system were no farther than the nearest star, Alpha Centauri, about 25 trillion miles away. For greater distances, a larger separation between the planet and its central star would be needed to distinguish the two.

Because the distances in space are vast, astronomers usually measure cosmic distances in terms of light-years rather than miles. A light-year is the distance light travels in a year, or about 6 trillion miles. In these terms, Alpha Centauri is four light-years away. At a distance of one hundred light-years, SIRTF could just detect the infrared radiation from a planet ten times more massive than Jupiter, if that planet's separation from its central star were twenty times greater than that of Jupiter from our sun. Such wide separations may indeed occur, and there are thousands of stars to search within one hundred light-years. An Earth-sized planet, unfortunately, would not emit sufficient infrared radiation and could not be detected in the foreseeable future.

Another promising method for finding planets is less direct. The central star of a planetary system should "wobble" slightly in response to the changing gravitational tugs of its orbiting planets.

Such shifts in a star's position are small. For example, the gravity of Jupiter causes the sun to move around in a circle whose diameter is only 1 million miles, the diameter of the sun itself. At the distance of Alpha Centrauri, four light-years away, this shift in position would appear as a shift in angle of only two-millionths of a degree, equivalent to the width of a penny seen from 250 miles away. Current visible-light telescopes on the ground cannot make out such tiny shifts in positions of a star, principally because the Earth's atmosphere blurs incoming light and makes all images of stars slightly fuzzy. However, the Hubble Space Telescope (HST), which was launched into orbit around the Earth in 1990, is above the atmosphere. When it is repaired, HST should have enough angular resolving power to sense the wobble of planet-bearing stars as far away as ten or twenty light-years, and the telescope will scrutinize a dozen candidates or so.

New "interferometric" space telescopes, which may be operational in a decade or two, could do even better. Interferometric telescopes can gauge changes of position highly accurately by measuring the changing overlap of light waves coming from slightly different directions. Such telescopes, now under development, should be able to infer the existence of planetary systems as far away as 300 light-years. Visible-light interferometric telescopes on the ground, less than a decade away, should also have significantly higher angular resolution than conventional ground-based telescopes. In the long term, astronomers hope to place an interferometric telescope on the moon.

The movement of central stars in response to their orbiting planets can also be detected through a phenomenon called the Doppler shift. The wavelengths of sound waves and light waves are altered by relative motion between sender and receiver. In sound, this alteration in wavelength is sensed as a shift in pitch. The whistle of an approaching train has a higher pitch than the same whistle of a train at rest; the whistle of a receding train has a lower pitch. In light, the analogue of pitch is color. Motion of an object toward the receiver causes the object's emitted light to shift to shorter wavelengths, toward the blue end of the spectrum; motion away causes the emitted wavelengths to lengthen, toward the red. As a central star moves around in a small circle, in response to an orbiting planet, its colors should shift toward the blue as it approaches Earth, then toward the red as it recedes, then toward the blue again, in a periodic pattern. Although the color shift should be tiny, corresponding perhaps to a speed of only ten or one hundred feet per second, it may be measured by instruments now in use and being developed. Such instruments have the ability, called spectral resolution, to measure colors with great accuracy.

Planets in the process of formation could be revealed by the study of gaseous protoplanetary disks in orbit about young stars. The Infrared Astronomical Satellite (IRAS) discovered a related phenomenon—disks of orbiting particles, which emit radiation at infrared and radio wavelengths. However, IRAS did not have sufficient angular resolution to see any details of the disks. Protoplanetary disks are believed to have diameters not much larger than our solar system, about one-thousandth of a light-year, and the nearest star forming (and planet forming) regions are about 300 light-years away. These numbers mean that an entire disk, as seen from the Earth, has an angular size of only 0.0002 degree, equivalent to the width of a penny observed from three miles away.

However, even these tiny sizes should be perceptible in the coming decade. The proposed Millimeter Array, working at radio wavelengths, will have the ability to resolve details as small as 0.00003 degree. The proposed Space Infrared Telescope Facility (SIRTF) will also be a major astronomical tool for the investigation of protoplanetary disks. This new infrared telescope will have ten times the angular resolution of its predecessor, IRAS, allowing it to discern details as small as 0.0002 degree in angle. The Space Infrared Telescope Facility will also have *a thousand to a million times* the sensitivity of IRAS and will consequently be able to detect cosmic emissions a thousand to a million times weaker. (With regard to sensitivity, SIRTF will be to IRAS what the great visible-light Hale telescope is to the naked eye.) By analysis of the infrared radiation emitted by protoplanetary disks, SIRTF may be able to determine their temperature, density, and composition—all useful information for the understanding of how planets form. Perhaps most important will be the surprises. The history of astronomy has shown that such enormous leaps in sensitivity of new instruments always lead to discoveries not even guessed at.

Weighing in at about five tons and with a length of about

sixteen feet, SIRTF will be launched in space and placed into orbit 65,000 miles above the Earth, or about eight Earth diameters away. In orbit, SIRTF will be able to detect radiation unable to penetrate the atmosphere, and it will be so distant that the Earth will rarely obstruct its view. Prior to the launch of SIRTF, the European Space Agency will place in orbit the Infrared Satellite Observatory, which will have capabilities intermediate between IRAS and SIRTF and which should also improve our knowledge of infrared phenomena in space.

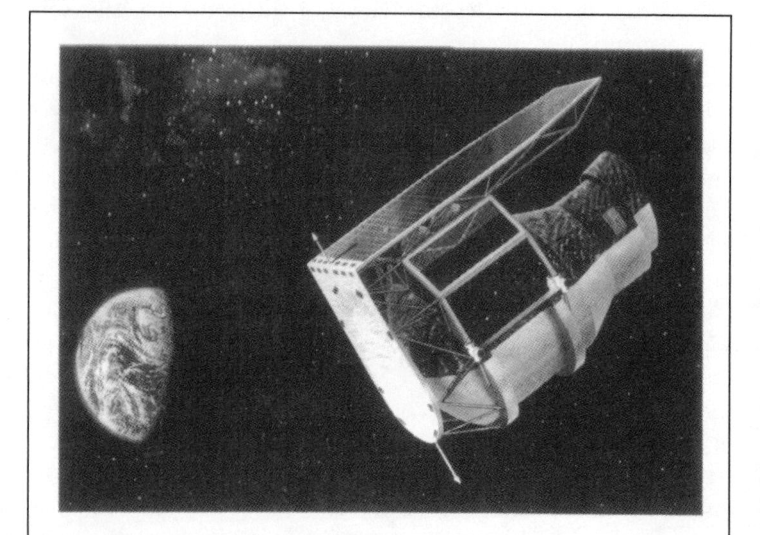

Fig. 5. *Artist's rendering of the proposed Space Infrared Telescope Facility (SIRTF), an Earth-orbiting telescope sensitive to infrared radiation.*

A new generation of telescopes on the ground should also have the needed angular resolution to study the structure of protoplanetary disks and other objects of very small angular size. One of these proposed new telescopes is called the 8-meter Infrared Optimized telescope, to be built on the summit of Mauna Kea, a dormant volcano on Hawaii Island.

Planned for the 1990s, the Infrared Optimized telescope (IRO) will be innovative in several respects. First, it will be sensitive to infrared radiation as well as visible light. Although Earth's atmosphere absorbs some wavelengths of infrared radiation, it allows others to pass through, and these will be monitored by IRO. Many visible-light telescopes built since 1970 respond to infrared radiation as well, but IRO is especially adapted to infrared radiation and will have a much improved performance for these radiations. Second, IRO will be larger than existing telescopes. Large telescopes now in operation typically have light-gathering mirrors of 160 inches (four meters) in diameter. For a doubling of its diameter, a telescope has four times the light-collecting area and thus can see objects four times as faint. As of 1991, the largest visible-light or infrared telescope in the world is the Soviet 236-inch telescope on Mount Pastuknov, in the Caucasus. The second largest is the Hale telescope on Mount Palomar in California, completed in 1949. Its light-gathering mirror is 200 inches in diameter. The Infrared Optimized telescope will have a mirror diameter of 315 inches (eight meters). With such light-gathering power, IRO should receive enough light from protoplanetary disks to split that light into its component wavelengths and thus identify the different kinds of molecules in the disk.

Third, the Infrared Optimized telescope, like other new infrared devices, will utilize the recently developed infrared-sensitive "array." Until a few years ago, each infrared camera had only a single detection element, analogous to a single rod or cone cell in the eye. An array is a mosaic of tens of thousands of detection elements, each one ten to one hundred times more sensitive than previous single elements. The net result is that the new arrays have a million times the sensitivity of previous infrared detectors.

Finally, the Infrared Optimized telescope will be equipped with a state-of-the-art mechanism called "adaptive optics," capable of correcting for the distorting effects of the Earth's atmosphere and thus allowing much sharper images of astronomical objects. Because of changes in temperature and moving pockets of air, the Earth's atmosphere is constantly shimmering. Such shimmering causes passing light rays to bend one way and then another, and any coherent images become smeared and fuzzy. In adaptive optics, the defocusing effect of the shifting atmosphere is compensated for by many motorized supports, placed behind the telescope's mirror. The supports reshape the mirror's surface every hundredth to a tenth of a second, in accordance with instructions from a computer. The computer, in turn, gets its orders by analyzing the image of a "guide star." If the atmosphere were perfectly smooth, the image of a star would be a single point of light. By analyzing the "fuzziness" of the guide star, the computer can infer the distorting effects of the atmosphere and instruct the motorized supports exactly how to reshape the mirror to bring the star's image and all objects near it back into sharp focus. If we think of the shifting atmosphere as a lens that is constantly changing its shape and focus, then adaptive optics constantly changes the telescope's mirror to mimic the atmospheric lens.

In the next decade, astronomers hope to install adaptive optics in a number of large, visible-light telescopes as well as in the Infrared Optimized telescope. Like the IRO, the new telescopes will be considerably larger than most existing telescopes, with mirrors ranging from 300 to 400 inches in diameter. The new telescopes are usually referred to as eight-meter to ten-meter telescopes. The first of these telescopes, the 396-inch (ten meter) Keck telescope, is under construction in Hawaii and will be operational in 1992. A European consortium is planning a group of four eight-meter telescopes, together called the Very Large Telescope, to be located in Chile. The new eight-meter and ten-meter telescopes may be able to see details in images as small as 3×10^{-5} degree, equivalent to the width of a penny observed from twenty miles away. If successful, adaptive optics will be able to confer some of the advantages of going into space, but at less cost.

We have mentioned angular and spectral resolution several times. In fact, an astronomical instrument is judged by several criteria: sensitivity, which is how well it can gather incoming light and respond to dim light; angular resolution, which is how well it can distinguish details in the image; and spectral resolution, which is how well it can separate incoming light into its component wavelengths. Instruments with high marks in these categories are necessary equipment for an observational astronomer. Necessary equipment for a theoretical astronomer includes a computer, plenty of paper and pencils, and a large wastebasket.

The Search for Extraterrestrial Intelligence

Are we alone in the universe? Few questions are more profound. Contact with other life forms would bridge the billions of years of independent starts and turns of life in the universe. Extraterrestrial contact would forever change the way we view our place in the cosmos.

It has taken 4.5 billion years for life on Earth to evolve to the realization that we live on a commonplace planet circling a commonplace star. The Infrared Astronomical Satellite found evidence of possible planetary systems. And we have discovered that space is littered with the carbon-based, organic materials needed for life as we know it. With radio telescopes we have now identified nearly one hundred different organic compounds in space, and with our hands we have held organic sludge delivered to us by meteorites. The ingredients are all there. What happened on Earth could have happened elsewhere.

For many years, astronomers have assumed that communication with extraterrestrial intelligence would best take place through electromagnetic waves, particularly radio waves. Ours is the first generation capable of such communication. The first modern search for intelligent, extraterrestrial radio broadcasts was conducted in 1960 by astronomer Frank Drake, then at the National Radio Astronomy Observatory in Green Bank, West Virginia. In this first effort, Drake pointed his antenna "ears" at two nearby stars and listened for two months. Since then, there have been about fifty such searches, mostly cheap and brief diversions of radio telescopes used for other purposes. In the past decade, we have mounted the beginnings of a sustained search, with sensitive detectors dedicated to the job of listening for possible intelligent signals from outer space. Specifically, what we are looking for is a narrow-wavelength-band broadcast coming from a single direction in space. If such a "signal" were found, we would then analyze it for encoded messages. Computer technology has contributed to our new receivers. The first searches for extraterrestrial broadcasts covered only 1,000 radio channels. The computer-automated receiving systems designed by Paul Horowitz of Harvard University now monitor 10 million channels simultaneously, a necessity when we are not sure what wavelengths the other fellows are broadcasting. By the end of the 1990s, receivers may have a *billion* channels.

In October of 1992, the National Aeronautics and Space Administration began a systematic search for intelligent extraterrestrial radio broadcasts using the Arecibo radio telescope.

The Life History of Stars

The Sun

In his masterwork the *Principia* (1687), Newton wrote that "those who consider the sun one of the fixed stars" may estimate the distance to a star by comparing its apparent brightness with that of the sun —in the same manner that the distance to a candle may be judged by comparing its brightness with that of an identical candle nearby. Newton then calculated that our nearest stars are about a million times farther than the sun, in good agreement with later measurements.

Implicit in Newton's argument is the assumption that stars are suns, much the same as our own sun. This is roughly correct. Indeed, some stars are identical to our sun, and most share its general features. Accordingly, we have learned about stars by studying the sun.

The source of heat and light on Earth has intrigued humankind through history. In the nineteenth century, scientists believed that the sun received its energy from chemical reactions, as if it were an enormous coal-burning furnace in the sky. It soon became apparent, however, that a sun running on chemical energy would have burned up its fuel in only 1,000 years. A better suggestion was that the sun was powered by gravitational energy, released from a slow sinking together of its mass, just as some grandfather clocks are powered by a slowly falling weight. However, even this mechanism could provide heat and light for only 100 million years—still much less than the 4.5 billion years reckoned to be the age of the Earth and solar system. Aware of this dilemma, the British astronomer Arthur Eddington proposed in 1920 that the sun and other stars are powered by nuclear energy. The nuclear reactions would take place at the center of the sun, where it is hottest and densest, and could provide the sun with energy for many billions of years.

The sun is a great ball of hot gas, about a million miles in diameter. According to modern theory, the density of the sun at its center is about one hundred times that of water, and the temperature is about 15 million degrees centigrade. Such high temperatures are needed to ram together subatomic particles violently enough to fuse them together and release nuclear energy. The liberated energy does two things. First, it maintains the heat within the material of the sun, providing sufficient pressure to resist the inward pull of gravity. (Without such pressure, the sun would collapse under its own weight.) Second, the liberated energy turns into radiation, which slowly makes its way to the solar surface and then out into space. Some of the sun's energy goes into churning up its surface and producing extremely energetic particles, magnetic fields, and an atmosphere of high temperature called the corona.

The solar corona mystifies astronomers. Easily seen during a solar eclipse as a brilliant crown of spikes poking out from the sun, the solar corona has been observed for centuries. But why is it so hot? And where does its energy come from? The temperature of the interior of the sun gradually decreases from 15 million degrees centigrade at its center to 6,000 degrees at its surface. These temperatures can be understood in terms of the balance of gravity and pressure forces within the sun and the luminosity and size of the sun. But there is no obvious reason why the gaseous material of the sun, after cooling off to 6,000 degrees at the solar surface, should quickly heat up again to several million degrees just above the surface, where it leaps into space to form the corona.

Some clues concerning the energy input to the solar corona have come from ultraviolet observations from space. Although unable to see patches of the sun as small as needed, perhaps fifty miles in size, the space observations have established the existence of myriad explosions and jets of gas within the corona. Astronomers believe that strong magnetic fields, by breaking apart and reconnecting, somehow pour energy into the sun's atmosphere. But the details of this picture are hazy. For example, how do the magnetic fields get *their* energy? And what is the role of magnetic fields in the curious eleven-year cycle of solar activity, first discovered in 1851 by the German apothecary and amateur astronomer Heinrich Schwabe?

The solar corona is a central player in much of the energetic activity of the sun. At its high temperature, the solar corona emits X-ray radiation and a stream of high-speed particles called the solar wind. The speed of the solar wind near the Earth's orbit averages about 300 miles per second. In addition to exuding a steady solar wind, the hot corona of the sun occasionally erupts in a violent burst of energy called a solar flare. A solar flare covering only one-thousandth of the sun's surface can outshine the rest of the sun in ultraviolet light and X-rays. An associated phenomenon is the dark patches on the sun, known as sunspots and first observed by Galileo in the early 1600s. Sunspots range in size from about 1,000 miles to about 30,000 miles. Individual sunspots come and go, lasting for a few hours to a few months. The overall number of sunspots, the solar corona, and the solar wind are interrelated and vary over time. When the number of sunspots is low, the solar corona extends far out into space near the solar equator but is very limited near the poles, while at times of high numbers of sunspots the corona is much more symmetrical.

Some scientists believe that the variable activity of the sun affects the climate on Earth, although the connection is not well understood. For example, historical records show that the number of sunspots and other related solar activity was much lower than normal during the period from 1645 to 1715, as first noted by German astronomer Gustav Sporer and British astronomer Edward Maunder about 1890. During this period of low solar activity, called the "Maunder minimum," the temperatures in Europe were unusually low. There is much more to learn about how, if at all, the sun's activity affects the climate on Earth.

In the last two decades, to astronomers' surprise, X-ray detectors and other instruments have found all these puzzling solar phenomena—coronas, strong magnetic fields, flares, stellar winds—in other stars as well as the sun.

The puzzle of coronas may be solved in the 1990s. Astronomers have proposed a new satellite called the Orbiting Solar Laboratory (OSL), dedicated to observing the sun. This new astronomical laboratory will carry a visible-light telescope, a telescope sensitive to ultraviolet light, and a telescope sensitive to X-rays. Complementing OSL will be a ground-based solar telescope called the Large Earth-Based Solar Telescope (LEST), to be located in the Canary Islands. The Large Earth-Based Solar Telescope will have very high angular resolution, for making images of detailed structures on the solar surface, and much higher precision than OSL for measuring the solar magnetic field. Among its missions, LEST will study the effects of magnetic fields on the transport of energy through the solar interior and will construct three-dimensional maps of the solar magnetic field.

Astronomers have little direct data from the interior of the sun. All the light we see comes from its surface. However, the surface contains clues about conditions far below. Of special importance are the movements of gases—a major new field of study called helioseismology. Vibrations of the solar interior, like earthquakes, carry information to the surface. Just as the intensity and intervals of terrestrial earthquakes and tremors tell us about conditions deep within the Earth, the gaseous vibrations on the surface of the sun inform us about the density, temperature, and rate of rotation in the deep interior. For example, recent analysis of observed solar vibrations has indicated that the deeper layers of the sun rotate at the same rate as the surface layers, contradicting theoretical expectations and baffling solar astronomers. For the most part, however, the results of helioseismology so far indicate that the solar interior conforms closely to theoretical preconceptions. Scientists are pleased when their theories are confirmed by experiment, but they are also often pleased, and always excited, when they find a contradiction.

Helioseismologists monitor the motions of the solar surface by relying on the Doppler shift, described in Chapter 1. As the solar surface vibrates, it moves first toward the Earth, then away, then toward again, and these to-and-fro motions are all precisely encoded in the pulsating variations in wavelength of emitted light. The motions of some solar vibrations repeat every five minutes and should be measured every minute or so for a period of several years in order to produce a good picture of the solar interior. Such long-term study requires many observatories spaced around the Earth, so that the sun will always be within view of at least one of them. To this end, astronomers are now building a network of at least ten solar observatories worldwide, called the Global Oscillation Network Group (GONG). Using the techniques of helioseismology, GONG will be able to measure the motions of 65,000 positions across the surface of the sun simultaneously.

A better knowledge of the interior of the sun could resolve another problem that has worried astronomers for years. The number of subatomic particles called neutrinos emitted by the sun is several times less than predicted. Neutrinos are electrically neutral particles that interact only very weakly with other particles. For many years, physicists have thought that neutrinos have no mass, that they are pure energy, like electromagnetic radiation. However, the properties of the neutrino, including its mass, are not well established experimentally.

In theory, neutrinos should be created in the nuclear reactions at the center of the sun. The rate of neutrino production has been carefully calculated by theorist John Bahcall of the Institute for Advanced Study and others, using our best theories of nuclear physics and knowledge of the conditions of temperature and density in the sun. Since neutrinos interact extremely weakly with other matter, almost every neutrino produced at the sun's center should escape to its surface and out into space. Thus, it should be relatively straightforward to compare theory to observation. In an experiment running

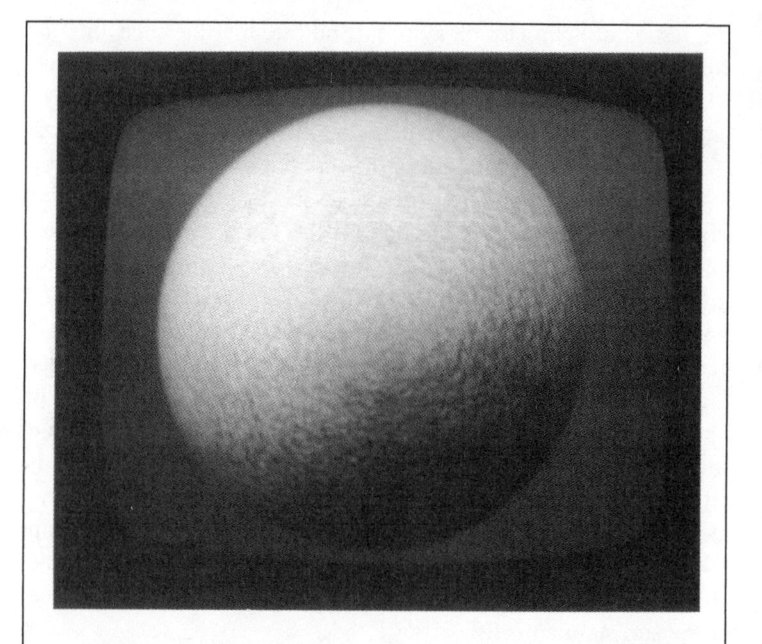

Fig. 6. Helioseismological image of the sun. Different shades indicate different speeds of the solar surface layers along our viewing direction. The center of the sun is approximately at rest. Regions with lighter shades than that of the center are moving toward us and regions with darker shades are moving away. The speckled pattern shows that neighboring regions can have very different motions, as the solar surface vibrates.

for the last twenty years, Raymond Davis and his colleagues at the University of Pennsylvania have measured and counted neutrinos from the sun and have found only about a third as many as predicted. This discrepancy has recently been confirmed in Japan. It is serious. Either the interior conditions of the sun are not what we think, or the neutrino has some property that allows it to change forms and avoid detection once emitted. The former explanation, if correct, could alter our theories of the structure of stars. The latter would have deep implications for our understanding of subatomic physics.

Planned for the 1990s are several new experiments to monitor neutrinos from the sun: a collaboration between Soviet and American scientists, called the Soviet-American Gallium Experiment, at the Baksan Laboratory in the former Soviet Union; the Gran Sasso Laboratory in Italy; and the Sudbury Neutrino Observatory, a collaboration of the United States, Canada, and Britain. These new experiments will differ from previous ones in that they will be sensitive to neutrinos of different energies, produced in different regions of the sun. The Sudbury Neutrino Observatory, moreover, should detect all forms of neutrinos. Even if neutrinos do change form, they will not be able to elude this new generation of neutrino detectors.

The Formation of Stars

It takes two things to form a star: matter, and a mechanism to compress the matter to high density. Matter is plentiful in space. It consists mainly of hydrogen gas, mixed with small amounts of other elements and small particles of dust. In some places the gas is rather smoothly distributed, and in others it is highly clumped. In a place of accumulated mass, gravity is stronger, and the gaseous clump can pull itself together, becoming even more dense. Thus gravity can provide the mechanism of compression. The dense, newborn stellar core might be a few tenths of a light-year in diameter, millions of times larger than a fully formed star. Several other forces combine with gravity to determine the behavior of the initial stellar core. Typically, the gas cloud will be spinning, and it will also be squeezed by magnetic forces, although the interplay of these additional factors is not completely understood. Also present within the cloud are heat and pressure, which resist compression. When the inward pull of gravity is sufficiently strong, the cloud will continue to contract and fall toward its center, producing heat from the release of gravitational energy, as does any contracting mass. (The heat eventually emerges as infrared radiation.) The collapsing gas cloud grows denser and hotter. In a rotating cloud, a disk of gas and dust the size of a solar system may form about the center. Eventually, the temperature at the gaseous center has risen to around 10 million degrees centigrade, enough to ignite nuclear reactions, and the cloud becomes a star.

Stars have masses ranging from 0.1 to 100 times the mass of our sun. Smaller masses never get hot enough at their centers to ignite nuclear reactions; larger masses blow themselves apart at formation by the outward force of their own radiation. The needed time for the birth of a star, as described above, varies with the mass of the star. According to theory, our sun required about 10 million years; a star of 0.1 times the mass of our sun would require about 100 million years, and a star of 100 times the mass of our sun would need only 10,000 years.

This theory of the formation of stars was given some support in the 1980s, when the Infrared Astronomical Satellite (IRAS) detected tens of thousands of stars in the process of formation. More specifically, IRAS detected dense stellar cores enshrouded in surrounding gas, during the early phase of collapse, before the nuclear reactions had begun. Our understanding of star formation was given another boost in the 1980s when radio telescopes made an unexpected discovery: streams of gas flowing outward in opposite directions from the vicinity of the embryonic star. Theorists argue that these gaseous streams may come from a planet-forming disk around the young star. However, their origin and role are not yet fully understood.

What keeps the star-forming clouds inflated for so long, without collapsing? What determines the mass of a new star and the number of stars formed of each size? What is the role of the gaseous disk that forms around a young star? What is the flow of gas in and out as a star is formed? To get at these questions requires continued theoretical calculations and new astronomical instruments. The new instruments must be able enough to distinguish clumps and filaments of gas as close together as about 3×10^{-5} degree in angle and must be able to measure wavelength shifts as small as 5×10^{-5} percent, corresponding to gas speeds as small as 0.1 mile per second.

Because star formation occurs within very dense clouds of gas, impenetrable to visible light, clues must be found in the trails of radio waves and infrared radiation. New instruments planned for the 1990s that will collectively satisfy these requirements are the Space Infrared Telescope Facility (SIRTF); the Stratospheric Observatory for Infrared Astronomy (SOFIA); the Infrared Optimzed telescope (IRO); the Millimeter Array (MMA); the Submillimeter Wavelength Telescope Array; and the Green Bank Telescope (GBT), in Green Bank, West Virgina. Each of these telescopes will make unique contributions. SIRTF will observe wavelengths that cannot penetrate the Earth's atmosphere. SOFIA will detect radiation from a variety of molecules and atoms that characterize the conditions in protoplanetary disks. The MMA and the IRO will make high-resolution studies of the protoplanetary disks and the outflowing gaseous streams from them.

The Millimeter Array, for example, will consist of forty different telescopes, each a dish about twenty-six feet in diameter. The telescopes will be electronically linked, so that they will be able to act in concert as one giant telescope. Capable of distinguishing two

Fig. 7. Artist's rendering of the proposed Millimeter Array (MMA), sensitive to radio waves.

distant objects separated by as little as 2×10^{-5} degree in angle, the MMA gets its name from the wavelengths of radio waves it is sensitive to, from about one millimeter to about ten millimeters (about 0.04 inch to about 0.4 inch). In this range of wavelengths, the MMA will have the highest angular resolving power of any telescope in the world. Sensitive to shorter wavelengths of radio waves will be the Submillimeter Wavelength Telescope Array, which will consist of a group of at least six antennas, each about twenty feet (six meters) in diameter.

The Life and Death of Stars

For centuries the stars have symbolized permanence, especially by contrast to earthly life. In Shelley's poem "Adonais" (1821) are the lines "The One remains, the many change and pass; / Heaven's light forever shines, earth's shadows fly." And in "Christabel" (1800), Coleridge wrote: "And constancy lives in realms above; / And life is thorny; and youth is vain."

Yet modern astronomers have found that nothing is permanent.

In the nineteenth century, when the distances to nearby stars were first directly measured, astronomers found that some stars appear brighter than others *at the same distance*. Thus, they concluded, all stars are not the same. Stars, like light bulbs, come in a range of luminosities. Stars also come in a range of colors. In the period 1911–1913, the Danish astronomer Ejnar Hertzsprung and independently the American astronomer Henry Norris Russell discovered a simple but crucial fact about stars. When Hertzsprung and Russell plotted nearby stars in a diagram with color along one axis and luminosity along another, they found that most stars lie along a diagonal band across the diagram. In other words, there is a definite relationship between color and luminosity: more luminous stars are bluer. If there were no such relation—if a star of a certain color could have a wide range of luminosities—then the stars plotted on a Hertzsprung-Russell diagram would be scattered all over the map. The diagonal band on which most stars lie is called the "main sequence." Stars at the bottom of the main sequence are red and dim; stars at the top are blue and very luminous. The principal factor that determines where a star lies on the main sequence is its mass. Heavier stars are bluer and more luminous; lighter stars are redder and less luminous.

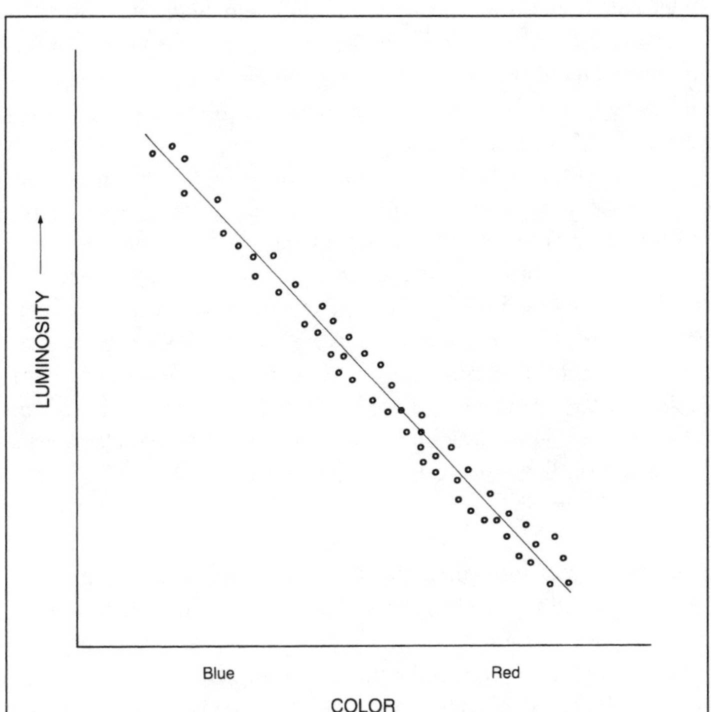

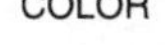

Fig. 8. Illustrative diagram of "main sequence" of stars. Each small circle represents an observed star. The horizontal position is the star's observed color; the vertical position is its observed luminosity. The fairly narrow diagonal band of stars across the diagram is called the main sequence.

Color is significant because color is closely related to temperature. Every hot body, including a star, emits electromagnetic radiation, and the color of that radiation is determined by the body's temperature. For example, a temperature of about 7,000 degrees centigrade produces violet light; a temperature 3,500 degrees produces red. (Temperatures higher than 7,000 degrees produce ultraviolet radiation, X-rays, and gamma rays; temperatures lower than 3,500 produce infrared and radio waves.) Observe the color of a star and you know its surface temperature. In effect, what Hertzsprung and Russell had discovered was a relationship between a star's luminosity and its surface temperature.

Hertzsprung and Russell found something else. A fraction of stars are peculiar; they do not share the relationship between luminosity and color of the majority of stars, called main-sequence stars. Some of these misfits are too luminous for their color, and some are too dim. From other evidence and theoretical considerations, we now believe that the misfits are stars at a later stage of their evolution. Indeed, stars are born, age, and eventually fade away or explode. Stars are not the "fixed" objects that Newton referred to, or the eternal lights of Shelley and Coleridge. Even Shakespeare's "constant" northern star will someday be gone.

Once a star has formed, it spends most of its active lifetime as a main-sequence star. In this stage, it burns its initial fuel, hydrogen, by fusing every group of four hydrogen atoms into a helium atom, releasing nuclear energy in the process. Then, after it has exhausted roughly 10 percent of its hydrogen, the star's central regions contract and its outer regions expand. The star continues to shine, however; it is powered by the gravitational energy released by its shrinking core. The stellar surface cools as it expands, causing the star to become much redder in color than a main-sequence star of the same luminosity. Such unusually red and large stars are called red giant stars, and they account for some of the misfits found by Hertzsprung and Russell. Eventually, the centers of sufficiently massive red giant stars become hot enough to rekindle new nuclear fuel, this time helium, the second lightest atom after hydrogen. Three helium atoms fuse together to make a carbon atom. Finally, after a sequence of nuclear reactions, fusing heavier and heavier atoms together, the core of the star is iron. Iron is the most barren of all elements. It will not yield energy either by fusing with other atoms or by fissioning into smaller atoms. Once the core of a star has matured into iron, and there exist no further resources of heat and pressure to counterbalance the inward pull of its own gravity, the star must collapse.

Our own sun has already lived about 5 billion years as a main-sequence star, and it will live another 5 billion years, quietly burning its hydrogen, before it swells into a red giant. Then, in the relatively brief period of about 100 million years, it will exhaust the rest of its nuclear fuel and collapse. More massive stars spend themselves more quickly and less massive stars more slowly. For example, a star of ten times the mass of our sun burns up its core of hydrogen gas and becomes a red giant in only about 30 million years. In general, more massive stars do everything more quickly than less massive ones.

How do we test our theories of stellar evolution? Clearly, we cannot wait long enough to see changes in individual stars. That requires millions or billions of years. What we can do, however, is observe many stars at different stages of evolution and from this infer the life story of a single star. In a similar way, botanists have unraveled the life cycle of California redwoods. Each tree lives many hundreds of years, much longer than a botanist, but by observing many trees, some just germinating, some sprouting first leaves, and others stretching grandly into old age, we can piece together the life history of a single redwood. Because new stars are continually born—on average, about ten new stars per year in our galaxy—in any large population of stars at any moment of time astronomers can find stars at every stage of evolution. The Hubble Space Telescope will study populations of stars millions of light-years away, far beyond our galaxy. In closely related measurements, the Space Infrared Telescope Facility will study the range of luminosities in newly formed stars.

A burned-out star may end its life in several ways. It can become a dense, dim star called a white dwarf or an even denser cold star called a neutron star. A white dwarf is 100 to 1,000 times smaller in diameter than a younger, main-sequence star of the same mass. A neutron star is 100,000 times smaller than a normal star and composed almost entirely of neutrons. Neutrons are subatomic particles that, together with protons, make up the centers of atoms. In a typical atom, the protons and neutrons reside in a dense central region, called the nucleus, making up more than 99.9 percent of the mass of the atom but only 0.0000000000001 percent of its volume. The vast majority of the atomic volume is taken up by the comparatively lightweight electrons, which orbit the nucleus at relatively large distances. In a neutron star, however, the neutrons are packed together side by side, solid. Imagine stripping the electrons off a sun's worth of atoms and cramming the nuclei together. You've got a neutron star. A neutron star has an unimaginable density: the mass of our sun compressed into a sphere ten miles across. Furthermore, that sphere spins very rapidly, between one and 1,000 revolutions per second; it anchors magnetic fields that are trillions of times stronger than Earth's; and it produces periodic pulses of intense radio waves. White dwarfs and neutron stars support themselves against further collapse by the resistance of their subatomic particles to being squeezed closer together. They can remain in such balance forever. Yet these massive spheres that were once shining stars have no source of energy, other than their energy of rotation, and so eventually grow dim and cold.

White dwarfs were identified about 1913 by Hertzsprung; in 1924 the British astronomer Arthur Eddington first estimated the size of a white dwarf star—the companion of Sirius, a star rather near to our sun. The first neutron star was identified in 1967 by Jocelyn Bell and Anthony Hewish of Cambridge University. Astonishing as it may seem, astrophysicists had predicted the characteristics of these strange objects before their discovery. Swiss-born astronomer Fritz Zwicky and German-born astronomer Walter Baade, working together in California, correctly forecast the existence and properties of neutron stars as early as 1934, only two years after the discovery of the neutron itself in terrestrial laboratories. Such accurate predictions testify to the validity of the assumption that the same physical laws found on Earth apply to distant parts of the universe.

Such predictions also testify to the power of theoretical calculations in astronomy. It may be worthwhile to digress for a moment to discuss the role of theory and observation in astronomy. In every

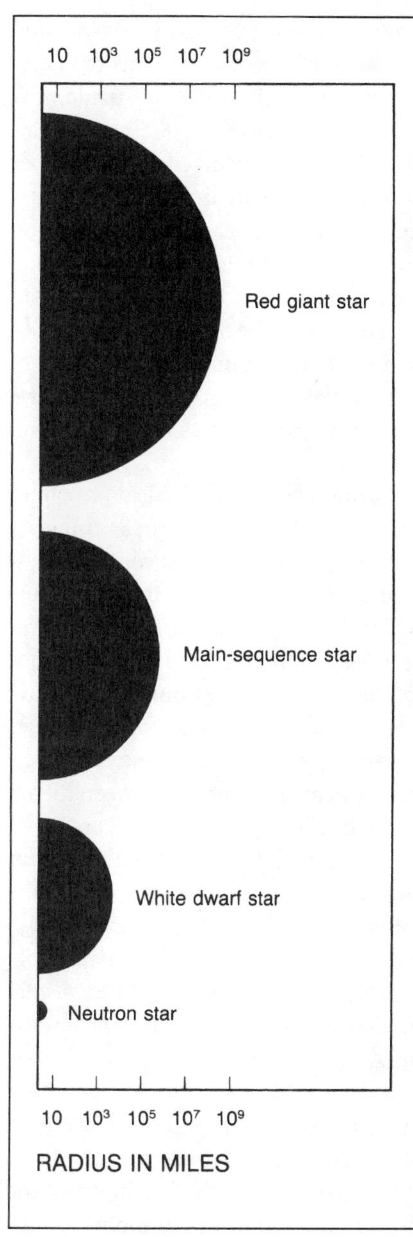

Fig. 9. *Approximate radii of the different types of stars. To cover the enormous range of sizes from neutron stars (radius about six miles) to red giants (radius about 200 million miles) a scale is used in which each unit to the right increases by a multiple of one hundred rather than by a constant additive amount.*

science, theory and experiment work together. Theory develops and unifies concepts, predicts the results of new experiments or observations, and interprets new observations after they are made. Experiment and observation tell us what is actually true in the world, outside our minds. Astronomy is somewhat different from other sciences in that the objects of study are usually light-years away, far beyond control. We cannot turn up the magnetic field in a neutron star and see how it will behave; we cannot rotate a galaxy this way and that to see it from different directions. In astronomy, we can only observe. In astronomy, people who collect data are called "observers," not experimentalists. Since in astronomy we cannot control what we observe, and we cannot isolate different pieces of what we observe, we must usually digest astronomical systems in their full complexity.

Many different phenomena and effects and physical principles are all tangled up together and often inseparable. For example, a particular intensity of radiation from gas around a neutron star might mean that the strong gravity of the star is affecting the gas in just the way that theory predicts, but that same intensity of radiation might also be produced by a messy clumping of the gas that has nothing to do with gravity and has no bearing on our elegant gravitational theory. Without being able to control the gas or get a closer look, there may be no way to distinguish between these two possibilities, or a dozen others. Thus, a clean comparison between theory and observation is much harder in astronomy than it is in many other sciences. In astronomy, it is difficult to show that a particular theory is irrefutably wrong or irrefutably right. This state of affairs give some theorists great comfort. It frustrates others. Nevertheless, even in astronomy, theory and observation have worked remarkably well. The observations of Hertzsprung and Russell provided vital information that went into our later theories of the structure of stars. The theoretical predictions of Zwicky and others allowed an interpretation of the curious observations of Bell and Hewish, who found an object that was apparently rotating once every second. Nothing could rotate that fast and hold together unless it had the small size of the hypothesized neutron star. We will see other examples of the successful interplay of theory and observation in astronomy.

Sallie Baliunas was born on February 23, 1953, in New York City. She majored in physics and astronomy at Villanova University and received her Ph.D. in astronomy from Harvard in 1980. Since then, she has been on the staff of the Harvard-Smithsonian Center for Astrophysics in Cambridge, Massachusetts. Baliunas's research has focused on the magnetic properties of the sun and other nearby stars, including the sun's eleven-year sunspot cycle and its counterpart in other stars. With her collaborators, Baliunas has developed a robot telescope, which is directed by a computer and can make automatic observations of stars.

Says Baliunas, "I have two recollections from my childhood: Sputnik and the wallpaper in my bedroom, which was printed with space-suited people rocketing toward Saturn. I believe that reaching for the stars is a fundamental

(CONTINUED FROM PREVIOUS PAGE)

ambition for humanity. Several broad areas in stellar astronomy are gaining advantage with new telescopes, computers, and electronic detectors. The mechanism that produces the changing magnetism of the sun and most stars is not fully known. Understanding the sun's varying magnetism is important for predicting its impact on our environment. With helioseismology and the GONG project, we will soon obtain the most detailed view to date of the sun's interior. Knowledge of the subsurface motions in the sun will be important in understanding the way magnetic fields are produced and in checking the validity of our models of the sun, both of which may bear upon explaining the low intensity of neutrnos observed from the sun. Finally, new instruments such as interferometers and spectrographs are beginning to search for evidence of planets orbiting other stars."

There are many unsolved puzzles with white dwarfs and neutron stars. What determines the initial rates of spin of these compact stars? What is the nature of the super dense matter at the center of neutron stars? The magnetic field of a neutron star, together with the spin of the star, accelerates electrons to high speeds and causes radio emission. Yet the emission fades away after 1 to 10 million years. Why? The new radio telescopes will attempt to answer these questions.

How are white dwarfs and neutron stars formed? Astronomers believe that all stars of a mass less than about eight times the mass of our sun will eventually collapse to form white dwarfs, after they have burned up their nuclear fuel. More massive, burned-out stars face a different end. They continue shrinking past the formation of a white dwarf, liberate an enormous amount of gravitational energy, and then explode in a brilliant display called a supernova. For a brief time, a single supernova can shine with the luminosity of 100 billion stars. (The word *nova*, meaning new, derives from the name used by astronomers many centuries ago. Historical astronomical records show that, from time to time, a "new star" suddenly appeared in the sky where there was nothing visible the night before. Novae and supernovae were called "guest stars" by the Chinese and first noted several centuries before Christ.) It is believed that a supernova often, and perhaps always, leaves behind a neutron star as a remnant. However, if the mass of the surviving neutron star is larger than about three times the mass of our sun, no amount of resistive pressure can stave off the overwhelming crush of gravity. In that case, the stellar carcass will collapse completely, forming a bizarre object called a black hole. In a black hole, which we will describe more in the next chapter, the gravity is so intense that not even a light ray can escape. It is the absence of any light emanating from such an object that gives it its name.

In early 1987, astronomers were handed a rare opportunity to test their theories of the evolution, collapse, and explosion of stars. A star exploded nearby, without warning, offering an unprecedented view of a supernova. (Here, "nearby" means in a neighboring galaxy.) By carefully monitoring the light from Supernova 1987A, as it is called, and by identifying from older photographs the star that blew up, astronomers have learned a great deal about the origin and nature of supernovae. The event was a triumph for theory as well as observation. The gamma rays from the radioactive decay of cobalt, the amount of nickel, the layering of silicon, oxygen, and other elements blown out by the shock wave—all were just as predicted. Also detected from Supernova 1987A were the elusive neutrinos. According to previous theories, neutrinos should be manufactured in great numbers during the formation of a neutron star. The neutrinos detected on Earth from Supernova 1987A not only confirmed the predicted temperatures and densities inside a supernova, but they also allowed physicists to place some limits on the properties of the neutrino. This is not the first time that answers to questions in theoretical physics have been found in the stars.

Supernova 1987A was discovered by accident. In the wee hours of February 24, 1987, on a mountain top in northern Chile, astronomer Ian Shelton from the University of Toronto was studying a photographic plate of the Large Magellanic Cloud, a small galaxy 170,000 light-years away. Shelton had just taken the picture with his small (10-inch) telescope when he noticed something odd. There was a very bright spot in the center of the picture. He went outside, peered up in disbelief, and confirmed his discovery by eye. There in the night sky he saw an intensely bright new star in the Large Magellanic Cloud where there had been none a few nights before. The supernova was also spotted by others, including Albert Jones, an amateur astronomer who had been examining the Large Magellanic Cloud from his backyard in New Zealand. Within hours, the news was reported to Brian Marsden, an astronomer at the Harvard-Smithsonian Center for Astrophysics in Cambridge. Marsden runs an official clearinghouse for astronomical data called the Central Bureau for Astronomical Telegrams. Word of Supernova 1987A spread rapidly around the world. Astronomers had detected other supernovae before. What threw them into such a feeding frenzy now was that Supernova 1987A was the closest supernova to have been observed for 383 years. Not since 1604 had a supernova been so near and so visible as to be seen by the naked eye. Here was a chance to study a supernova at close range, relatively speaking.

Still on February 24, and as soon as he got wind of the discovery, Harvard astronomer Robert Kirshner telephoned NASA and made arrangements for an orbiting astronomical satellite, the International Ultraviolet Explorer, to immediately change its scheduled program and point toward Supernova 1987A. And thus began scrutiny of the new supernova with an array of high-tech satellites, telescopes, and computers—all of which would have seemed like the devil's magic to the last person who witnessed a supernova with his naked eye, German astronomer Johannes Kepler, in 1604.

Why do some massive stars become supernovae and others black holes—or both? Neither of these endpoints of stellar evolution is well understood. According to estimates, a sizable fraction of stars should have initial masses larger than ten times that of our sun, qualifying them to become black holes. Yet only several black hole candidates have been identified, out of the billions of stars in our galaxy. Evidently, most massive stars manage to lose a large fraction of their mass before exhausting their nuclear fuel, or else black holes are even more difficult to find than believed. With supernovae, the dilemma is reversed. Nature has no trouble making supernovae, but theoretical astronomers do. So far, computer simulations of burned-out stars have been unable to coax the hypothetical star to properly explode. While it is easy to get the star to collapse, it is hard to get it to splash and rebound. Something, it seems, is missing in the instructions to the computer.

Ninety-nine percent of the explosive energy release in observed supernovae comes out in neutrinos; the remaining energy appears in the form of kinetic energy of expansion (energy of motion) and in X-rays and gamma rays. Study of these emissions may answer the question of what causes the explosion. Neutrino detectors now in existence and those being built will be on the alert to study neutrinos from future supernovae, continuing the successes with Supernova 1987A. Gamma-ray observations will also reveal the various chemical elements formed in the supernova explosion. Just as the precise wavelengths of radio and infrared emission uniquely identify the radiating molecules, so the wavelengths of gamma-ray emission identify the radiating atomic nuclei. The central nucleus of each kind of atom emits gamma rays with a wavelength unique to that atom. In particular, cobalt, nickel, iron, and titanium are formed in supernova explosions, and the analysis of their abundances and various forms can tell much about the workings of the supernova.

In April 1991 astronomers launched an Earth-orbiting gamma-ray detector, called the Gamma Ray Observatory (GRO), which has more sensitivity than any previous gamma-ray detector in astronomy and which will study supernovae, among other projects. This new orbiting observatory weighs fourteen tons and extends seventy feet between the tips of its solar-power arrays. By the end of the 1990s, astronomers may be ready to launch an even more advanced gamma-ray detector, the Nuclear Astrophysics Explorer (NAE), which will have the sensitivity to measure gamma rays from supernovae 100 million light-years away. Since a supernova happens at most once in any star's lifetime, it is necessary to monitor huge volumes of space in order to catch some star in the process of exploding. NAE will be able to measure wavelengths of gamma rays to an accuracy of about 0.1 percent, thus providing a very good identification of the atoms responsible for the emission.

Another major astronomical observatory planned for the 1990s is the Advanced X-ray Astrophysics Facility (AXAF). This new facility in space will be the successor to the Einstein X-ray Observatory, which orbited the Earth from 1978 to 1981. An orbiting cylinder forty-six feet long and thirteen feet in diameter, AXAF will weigh about thirteen tons. Like the Einstein Observatory, AXAF will have special mirrors able to focus X-rays and make images of objects through X-ray light. Since head-on X-rays do not reflect from conventional mirrors, AXAF's six gold-coated mirrors and detectors are so arranged that the incoming X-rays glance off them at angles of only a few degrees and then are recorded. The Advanced X-ray Astrophysics Facility will have ten times the angular resolution of the Einstein Observatory, allowing it to distinguish features only 0.0001 degree apart; AXAF will also have one hundred times the sensitivity of Einstein and 1,000 times the spectral resolution. Among its many scientific missions, AXAF will analyze the X-rays emitted by supernovae.

The Advanced X-ray Astrophysics Facility, like the Hubble Space Telescope, the Space Infrared Telescope Facility, and the Gamma Ray Observatory, is one of a new breed of astronomical instruments designed to operate in space, high above the atmosphere. The Earth's atmosphere, while usually agreeable to our noses, is a headache to astronomers. As mentioned earlier, images of astronomical objects are blurred as the light travels through the clumpy and turbulent air around the Earth. Moreover, many wavelengths of radiation get absorbed by the air and never reach the ground.

It was in 1923 that the German rocket pioneer Hermann Julius Oberth first pointed out that space is where telescopes ought to be. But Oberth's idea had to wait for technology to catch up with it. The first astronomical observations in space were made in the late 1940s, using captured German V-2 rockets capable of poking above the atmosphere for a few minutes. The first orbiting astronomical satellites, with the precious advantages of stability and long lifetime, were operated by the National Aeronautics and Space Administration

Fig. 10. Artist's rendering of the proposed Advanced X-ray Astrophysics Facility (AXAF), an Earth-orbiting telescope sensitive to X-rays.

in the 1960s. At about the same time, microchips landed on the scene, allowing miniaturized computers to be stowed aboard the satellites. Such computers control the program of observations of the telescopes and direct them to point in different directions at different times. After an orbiting telescope has taken its images, computers and other high-technology devices relay those images to the ground. There are no film drops. The pictures from the telescope are recorded in the form of electrical impulses, digitized (converted into bits of information represented by patterns of zeros and ones), and then transmitted by radio to other relay satellites or directly to Earth. Supernova 1987A, for example, was monitored by the orbiting International Ultraviolet Explorer, after hasty reprogramming by radio turned its electronic eyes toward the unexpected blast. Such quick reactions are necessary for supernovae, which flare up from nothing in only a few hours and never announce their intentions in advance.

Supernovae play a vital role in the life cycle of stars. The debris from stellar explosions spreads out into space and adds new ingredients to the gas between stars from which new stars form. Thus, supernovae are beginnings as well as ends. Theoretical calculations suggest that essentially all of the chemical elements except hydrogen and helium—the two lightest elements—were originally manufactured by the nuclear fusions within stars. Before the first stars, some 10 billion years ago, there was only hydrogen and helium. The vast majority of the one hundred chemical elements, including oxygen and carbon and other elements that earthly life depends on, were synthesized in stars and blown into space. Some of this "seeding" occurs during the red-giant phase, as a star sheds its surface layers, and some of it occurs in the wind of particles that flow from the hot stellar atmosphere. The rest happens in supernovae explosions. Later generations of stars, such as our sun, are born from the gas enriched by these new elements. Indeed, the different generations of stars are distinguished principally by their chemical composition. Stars formed entirely from hydrogen and helium are "first generation." Stars formed from heavier elements as well are second and later generations. The gas between stars connects the generations. It takes from the old stars and gives to the new.

Examination of the X-rays and gamma rays emitted by supernova remnants will help identify the various kinds of atoms dispersed in supernovae. Such a task will be on the dockets of the Gamma Ray Observatory, the Nuclear Astrophysics Explorer, and the Advanced X-ray Astrophysics Facility. Chemical elements manufactured and spewed out into space by supernovae can also be identified by their infrared emissions. The proposed Stratospheric Observatory for Infrared Astronomy, quickly deployable by an aircraft, will have the flexibility to study supernovae debris on short notice. And the high sensitivity of the Space Infrared Telescope Facility will permit more leisurely infrared measurements of supernovae out to thirty million light-years from Earth.

Old and new stars are connected in one final way. Shock waves emanating outward from supernovae compress the surrounding interstellar gas. Such compression may trigger the birth of new stars. Furthermore, supernova explosions energize the gas between stars. The radiation and expanding debris from supernovae heat large volumes of gas to 1 million degrees. At such temperatures gas radiates in the ultraviolet and X-ray region of the electromagnetic spectrum and will be studied by the Advanced X-ray Astrophysics Facility and two proposed ultraviolet instruments, the Far Ultraviolet Spectroscopy Explorer and the Extreme Ultraviolet Explorer. The altered contours of disturbed gas could reveal the causes and conditions under which some stars form.

The Life History of Galaxies

The Discovery of Galaxies

Among the ancient mysteries of the heavens were the misty patches, or "nebulae," noted by Hipparchus in the second century B.C. Nebulae were too distant to be atmospheric clouds, yet too diffuse for single stars. What were they? In 1610, with his new gadget the telescope, Galileo was delighted to find that "the [misty patches] which have been called by every one of the astronomers up to this day nebulous, are groups of small stars set thick together in a wonderful way, although each one of them . . . escapes our sight."

One nebula is the faint, milky band of light that sweeps across the sky at night. It was called the "river of heaven" in ancient Mesopotamia and is known as the "Milky Way" today. In 1785 the British astronomer and musician William Herschel measured the density of starlight in different directions and found that the Milky Way is shaped like a grindstone. That grindstone contains about 100 billion stars, orbiting about the center of the system to form a flattened disk. Such a congregation of stars is called a galaxy. The Milky Way is our own galaxy. (The word "galaxy" stems from the Greek word *galaxias*, meaning milk.) We now know that the universe is filled with galaxies, each containing gas and billions of stars. It takes our sun about 200 million years to complete one orbit about the center of the Milky Way. Galaxies come in a variety of shapes. Some are nearly spherical, while others, like the Milky Way, are flattened disks with a bulge in the middle.

Fig. 11a. Barred spiral galaxy NGC 3992.

Fig. 11b. Spiral galaxy NGC 4565.

For a century and a half after Herschel's work, astronomers had little idea of the size of the Milky Way or even whether other galaxies existed. The problem was in measuring the distance to stars. When we gaze at the sky, we readily perceive width and breadth, but not depth—as if we were looking at a photograph with no clues of the actual sizes of the objects in it. Furthermore, the apparent brightness of a star is not necessarily a good indication of its distance, since stars come in a wide range of luminosities, as discussed in the previous chapter. A star that appears dim could be of average luminosity and far away, but it could also be of very low luminosity and nearby.

A breakthrough in astronomical distance determinations came in 1912, when Henrietta Leavitt of the Harvard College Observatory discovered a remarkable feature of certain stars called Cepheid variables. Cepheid variables, first identified in the eighteenth century, oscillate in brightness, growing dim, then bright, then dim again, in regular cycles. (We now know that the variations are caused by periodic expansions and contractions of the stellar surface.) Regular oscillations in brightness are the fingerprints of these peculiar stars. After studying many photographic plates, Leavitt found a definite relationship between the cycle time of Cepheid stars and their luminosities. Such a relationship can be calibrated for nearby Cepheids, where the distances are known. Then it can be used to gauge the distances to remote Cepheid stars. For example, astronomers might observe that a particular Cepheid star of unknown distance has a light-cycle time of ten days. According to Leavitt's relationship, this cycle time corresponds to a luminosity of about 2,000 times that of our sun. Combining this luminosity with the observed brightness of the star, astronomers could then determine its distance, just as the distance to a light bulb may be determined from its wattage and observed brightness. If an astronomer wants to know the distance to a globular cluster or nebula or galaxy, she simply has to find a Cepheid star inside it. Cepheid stars are the road signs of space.

It is impossible to overestimate the importance of Leavitt's work, although Leavitt herself received few honors during her lifetime. Distance measurements are central to astronomy. Henrietta Swan Leavitt was born on July 4, 1868, in Lancaster, Massachusetts.

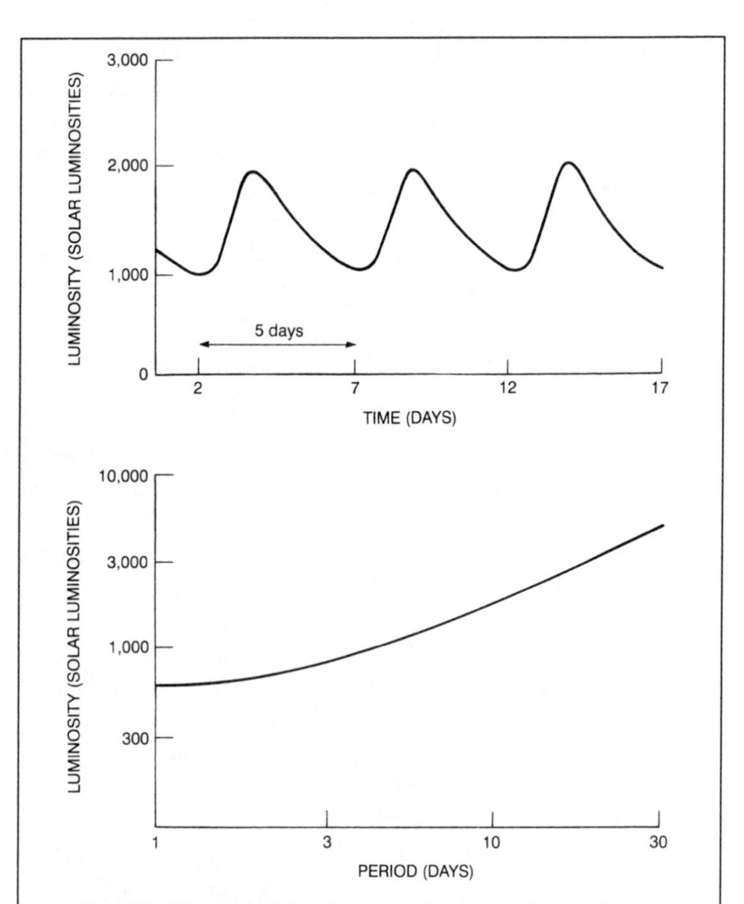

Fig. 12. The period luminosity relation for Cepheid variable stars. The top curve shows the variation in time for the luminosity of a typical Cepheid. The luminosity of this particular star varies between about 1,000 and 2,000 times that of the sun, on a cycle of about five days. The bottom curve shows how the average luminosity of Cepheid stars varies with their cycle time, or period. The particular Cepheid star shown in the top curve corresponds to one point in the bottom curve.

She was the daughter of a Congregational minister and throughout her life held to the strict Puritan virtues of her parents. Leavitt received an A.B. in astronomy from Radcliffe College in 1892 and soon became a volunteer research assistant at the nearby Harvard College Observatory. In 1902 she received a permanent, paid position at the Observatory and became chief of the astronomical photography department. Her work involved comparing numerous photographic plates, taken in time sequence, and determining which stars had varied in brightness from night to night and by how much. Such work demanded extreme patience and skill. In her career, Leavitt discovered and analyzed 2,400 variable stars. Leavitt was among a group of women, including Williamina Fleming and Annie Jump Cannon, who had been hired by the director of the Harvard College Observatory, Edward Pickering, to analyze photographic plates of stars. At the turn of the century, photographic work was just making its debut as a valuable tool in astronomy, and Leavitt,

Henrietta Leavitt (1868–1921).

Cannon, and Fleming were among the first to use this new tool.

In 1918 Harlow Shapley, another seminal American astronomer who was soon to succeed Pickering as director of the Harvard College Observatory, set to work measuring the distances to 230 Cepheid stars all over the Milky Way. When Shapley finished, he had made a detailed chart of the Milky Way, estimating its diameter as some 300,000 light-years across. The modern value is about 100,000 light-years. Our own sun, a member of the Milky Way, is about two-thirds the way out from its center. Shapley's work provided the first reasonably accurate measure of the size and shape of our galaxy.

But the question remained: What and where were the other nebulae? Were they clusters of stars in our galaxy, the Milky Way, or were they other galaxies themselves? Astronomers fiercely debated this question. At least some of the nebulae were within our own galaxy. In 1924 the American astronomer Edwin Hubble, with a telescope at the Mount Wilson Observatory in California, found a Cepheid star in the Andromeda nebula and measured its distance. Andromeda was a congregation of stars about 2 million light-years away, far beyond our galaxy. Andromeda was undeniably another galaxy. Edwin Hubble became the father of extragalactic astronomy.

Before long, astronomers had determined that many of the faint nebulae were, in fact, entire galaxies. With this knowledge came a new and larger view of the cosmos. Attention turned to galaxies. On average, galaxies are separated by about 10 million light-years, or about one hundred times the diameter of one galaxy. Thus, to a giant cosmic being, space would appear as a mostly empty sea, with isolated islands of stars, the galaxies, scattered here and there.

From another perspective, however, galaxies are far closer together than stars. Individual stars in a galaxy are separated by an average of ten light-years, or 100 *million* times the diameter of one star. Thus, a patchwork quilt of a galaxy, in which each patch was the size of a star, would have 100 million patches between stars. But a quilt of the universe, in which each patch was the size of a galaxy, would have only one hundred patches between galaxies.

Dissecting the anatomy of galaxies has been a major preoccupation of modern astronomers. Why are some galaxies mostly unformed gas and dust, while others are mostly stars? What determines the shape of a galaxy? Why are some galaxies nearly spherical, while others are flattened disks? In spiral galaxies, like our own, does the central bulge form first and then the disk, or vice versa? What determines the odd features of some galaxies—the rings and warps and bars? Were these features built in at the beginning, or did they form later, as the result of gravitational forces within the galaxy? Or perhaps they were fashioned by the close encounter or merger with another galaxy.

A stunning advance in galactic anatomy has been the realization that most galaxies must have enormous halos of invisible matter, called dark matter. In the mid-1970s, Jeremiah Ostriker and James Peebles of Princeton and Amos Yahil of the State University of New York at Stony Brook combined data from the orbital motions of a wide variety of astronomical systems—ranging from pairs of orbiting galaxies to large groups of orbiting galaxies—and arrived at the conclusion that there must be at least ten times more mass in these systems than was visible. A similar analysis was done independently by J. Einasto, A. Kaasik, and E. Saar of Estonia. The existence of dark matter was clinched in 1978 by the work of Vera Rubin and colleagues at the Carnegie Institution of Washington and, independently, by Albert Bosma of the University of Groningen. By studying the speed of orbiting gas in nearby galaxies, these researchers found clear observational evidence for massive, invisible (nonluminous) halos around individual galaxies. The visible mass alone could not propel the orbiting gas at nearly the speeds that were observed.

By now, almost all scientists accept the reality of dark matter. Dark matter in the halos of galaxies is needed to explain the stability of galaxies, the rotational speeds of stars and gas in galaxies, and the ease with which galaxies appear to collide and merge with each other. Photographs of galaxies, like pictures of icebergs, give no hint of these massive attachments. In the next chapter, dark matter will be discussed in more depth.

Astronomers have also learned something of the shapes of galaxies by measuring the *location* of orbiting hydrogen gas. Such gas can be found in regions uninhabited by stars, and it emits distinctive radio waves at a wavelength of twenty-one centimeters. Like radioactive iodine administered to patients before an X-ray, hydrogen gas acts as a tracer. And the gas is already in place. More recently, astronomers have found carbon monoxide gas in hundreds of galaxies. This gas, which can be detected at greater distances than can hydrogen gas, also emits distinctive radio waves and has revealed galactic twists and turns previously unseen. New radio telescopes, such as the Millimeter Array, will be able to form detailed images of distant galaxies through the detection of carbon monoxide.

The Evolution of Galaxies

For many years after their discovery, galaxies were assumed to be fixed and unchanging. The apparent brightnesses of galaxies were used to gauge their distances, as if galaxies were light bulbs of known wattage. Beginning in the 1950s, this picture of tranquillity began to change. New radio telescopes found some galaxies that radiated intense radio waves from great streams of gas emerging from the galactic centers. From an analysis of such radiation, astronomers infer that the gas streams are traveling at close to the speed of light. Since these gas streams are observed to be only 1 million light-years long at most, they could not have been flowing for much more than a million years, a relatively short time by astronomical standards. Clearly, something is changing in these "radio galaxies." You don't see jets gush from dead geysers.

Radio galaxies were the first galaxies to show clear signs of violent behavior. Their discovery was tied to the detection of cosmic radio waves, the first form of nonvisible electromagnetic radiation to be "seen" with astronomical instruments. Radio astronomy was born in 1931, when Karl Janksy, an American engineer at the Bell Telephone Laboratories, built a rotating radio antenna and discovered a steady static from outer space. In the following fifteen years, the only astronomer *in the world* to follow up on Janksy's accidental discovery was Grote Reber, an electronics engineer, radio ham, and amateur astronomer.

Reber was born in Wheaton, Illinois, in 1911. In 1936, in his backyard in Wheaton, the twenty-five-year-old Reber constructed a large metallic dish, thirty-one feet in diameter, supported by a tower of wooden two-by-fours. This was the first radio telescope, an antenna especially designed to receive radio waves from the cosmos. By 1942, Reber had made the first radio maps of the Milky Way. It is believed that until the mid-1940s, Reber was the only radio astronomer. In the late 1940s, other radio telescopes were built in Australia, England, the Netherlands, and the United States, and radio astronomy became established. The discovery in the 1950s of radio-emitting streams of gas in certain galaxies was the first announcement of galactic activity and evolution.

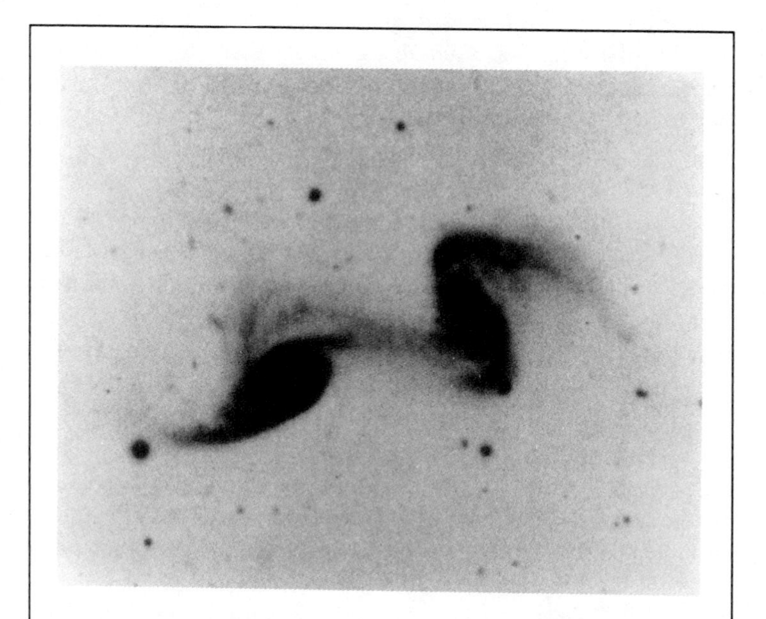

Fig. 13. Photo of two galaxies having a close encounter and distorting each other by their mutual gravity. The two galaxies are NGC 5426 *and* NGC 5427.

By the 1970s, astronomers realized that *all* galaxies should evolve. An astronomer from New Zealand working at the University of Texas, Beatrice Tinsley, pointed out that galaxies are made of stars. Stars age and change. Stars also alter the chemical composition of interstellar gas, in a one-way process that builds heavier and heavier atoms. Thus, the chemical composition, color, and luminosity of a galaxy should all change in time.

Galaxies can also change by interacting with other galaxies in their vicinity. Under the influence of their mutual gravitational attraction, galaxies cluster together. In a group of galaxies, where galaxies are closer together than normal, individual galaxies can become trapped in one another's outer envelope of stars, spiral together, and produce a new, composite galaxy of different properties than either of its progenitors. In photographs of some groups of galaxies, the central galaxies appear oddly entangled, suggesting a merger under way. In other pictures, there is only a single galaxy, but with peculiar tails, ripples, and other oddities suggestive of a disastrous encounter with a companion.

Finding direct evidence for evolution is much harder for galaxies than for stars. Stars are continually being born, so that any large volume of space contains stars at every stage of evolution. But most galaxies were probably born at about the same time in the past. Thus, at any moment of time—today, for example—most galaxies may be about the same age. How then can we observe galaxies at different periods of their lives?

The trick is light. Light travels at a finite speed, 186,000 miles per second, and the distances in space are large. When we take a picture today of the Andromeda galaxy, 2 million light-years away, we see that galaxy as it was 2 million years ago; it took 2 million years for the light to travel from there to here. When we look at a galaxy in the Virgo cluster of galaxies, 50 million light-years away, we see light that was emitted 50 million years ago. Hence, *looking deeper into space is looking further back in time*. Telescopes are time machines. With telescopes, we can see galaxies at increasingly earlier stages of evolution.

Unfortunately, the light from distant galaxies is faint. To detect such feeble light, astronomers need large telescopes. Until a couple

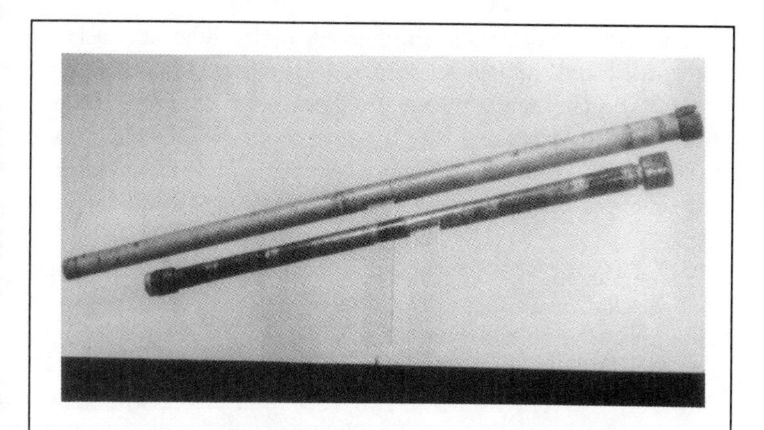

Fig. 14. Two of Galileo's original telescopes from 1610, each a few feet long, preserved at the Museum of History of Science in Florence.

of decades ago, there were few large telescopes and relatively primitive instruments connected to those telescopes. Few galaxies were recorded beyond a billion light-years away. Beginning in the early 1970s, a number of large new telescopes were built, including the telescopes at Mount Hopkins and Kitt Peak, near Tucson, Arizona, and at Cerro Tololo, Chile.

Even more important, there was an explosion of new instruments and technology for gathering and recording dim light. Photographic plates were replaced with computer-driven digital detectors, which can record ten to one hundred times as much of the incoming light and which translate the light into electrical signals, rather than darkened grains on a photographic plate. Electrical signals are easy to work with. They can be digitized, stored in the magnets of computers, and manipulated at a later date. For example, if the image of a galaxy is scrambled by light from a foreground galaxy of known properties, the computer can electronically subtract out the light of the second galaxy and reconstruct a clear image of the first.

In the present decade astronomers are building several large, visible-light and infrared telescopes, with diameters ranging from about 300 inches (eight meters) to about 400 inches (ten meters). With the new generation of large telescopes and their increased ability to see faint light, astronomers hope to see more distant galaxies at a much earlier stage of evolution than any galaxies previously seen.

For two decades away, astronomers are dreaming of a new orbiting telescope in space, the Large Space Telescope (LST). The successor to the Hubble Space Telescope, LST will have a mirror six meters in diameter and ten times the angular resolution of HST at its shorter wavelengths. The Large Space Telescope will be sensitive to radiation all the way from infrared to ultraviolet wavelengths and will study extremely distant galaxies. In addition, LST will probe star-forming regions, search for other planets, and analyze the gas between stars.

What types of stars inhabit young galaxies? While individual stars are born and die, how does the bulk population of stars in a galaxy age in time? Does the shape of a galaxy change in time, or is it completely determined when a galaxy first forms? How does the total luminosity of a galaxy change in time? How do the neighboring galaxies in groups and clusters of galaxies affect each other? These are all critical questions in extragalactic astronomy.

Astronomers believe that many, if not all, galaxies went through an extremely energetic early phase of evolution in which almost all of their energy was produced in their centers. This belief is based on the discovery of quasars in the 1960s. Quasars resemble single stars on photographic plates, yet have luminosities exceeding those of entire galaxies. Quasars were discovered accidentally in 1963 by Maarten Schmidt, of the Palomar Observatory at the California Institute of Technology. Schmidt realized that certain features of the radiation from these objects (their colors and so-called redshifts, to be discussed in the next chapter) suggested that they are at vast distances from Earth. Only an extremely luminous object could appear bright and yet be so far away. Furthermore, the very small size of quasars indicates that an enormous amount of energy is produced in a tiny volume of space, perhaps as small as our solar system. Quasars constitute the most energetic astronomical objects in the universe, and they came as a complete surprise. No one predicted quasars.

It is significant that most quasars have been found only far away. There are few quasars nearby. Since distance translates into time in astronomy, we can infer that most quasars lived and died in the distant past. They are the dinosaurs of the cosmos. Surrounding each of the closer quasars, astronomers have discerned the faint outlines of a galaxy. For these reasons scientists theorize that quasars constituted the central regions of galaxies at a very early stage of their evolution. The new generation of large, visible-light, and infrared telescopes and the already launched Hubble Space Telescope may be able to resolve the weak light of many more infant galaxies containing quasars and thus provide critical information on the connection between quasars and galaxies. The big question now is why some distant galaxies contain quasars and some do not.

New infrared telescopes, such as the Space Infrared Telescope Facility (SIRTF) and the ground-based infrared telescopes, will also play important roles in quasar research. Great dust clouds apparently surround many quasars, obscuring their visible light. When the energy of a quasar is absorbed by the dust cloud, it is re-emitted as infrared radiation. In the 1980s, the Infrared Astronomical Satellite discovered certain extremely luminous galaxies that emit 90 percent or more of their energy as infrared radiation and apparently harbor quasars at their centers. Furthermore, many such galaxies appear to be colliding with other galaxies. Could collisions of galaxies give birth to quasars, or refuel them? With its much greater sensitivity, SIRTF should be able to study the nature and evolution of these curious "infrared galaxies." If a big fraction of quasars are produced by collisions of galaxies, SIRTF will find out. Finally, new radio telescopes with extremely high angular resolution, particularly the Very Long Baseline Array, should be able to make radio-wave images of quasars themselves. Indeed, quasars were first discovered as a result of their intense emission of radio waves.

Aside from a possible quasar phenomenon at a young galaxy's center, what should a young galaxy look like? Astronomers are not sure. Galaxies were probably formed about 10 to 20 billion years ago, perhaps a few hundred million years after the beginning of the universe itself. As will be discussed shortly, astronomers believe that galaxies are born much the same way that stars are: by the compression of a cloud of gas. If the compression of gas on a large scale, to form a galaxy, is accompanied by the simultaneous compression of gas on small scales, to form stars, then the birth of a galaxy should be accompanied by the birth of many stars within the galaxy. Stars in the process of formation emit infrared radiation and should be observable with infrared telescopes. The Space Infrared Telescope Facility and the Hubble Space Telescope, which will be equipped with infrared detectors, will have the sensitivity to see such infrared emissions from primeval galaxies. The proposed Infrared Optimized telescope will be sufficiently sensitive to break up dim infrared light into its component wavelengths and thus identify the chemical elements in young galaxies. Astronomers have found evidence for star formation in extremely distant radio galaxies and a surprising association between the region of star formation and the gaseous jets emanating from the centers of the galaxies. It is possible that the high-speed jets compress the surrounding gas and thus trigger the formation of stars.

The Power Source of Quasars and Active Galaxies

Where do quasars get their great energy? Indeed, questions of energy arise for a variety of highly luminous galaxies found since the discovery of radio galaxies in the 1950s. Many of these "active galaxies" emit huge amounts of X-rays and infrared radiation as well as radio waves and visible light, and the energy appears to originate from an extremely compact region at the center of the galaxy, possibly the same size as a quasar. Many active galaxies propel great streams of gas from their centers.

Most of the energy for these objects certainly cannot come from stars. Stars radiate predominantly visible light. Furthermore, stars live on nuclear energy, which, with an efficiency of only about 0.5 percent at converting matter into energy, is not efficient enough to explain the huge energy requirements of quasars and active galaxies. Finally, the stars in a galaxy are scattered about, while the energy of active galaxies is produced in a highly concentrated region at their centers. Even if a sufficient number of stars were somehow crammed into such a small volume, the resulting stellar system would be so dense that the stars would quickly collide with each other and coalesce into a single massive object.

For these reasons, most astronomers believe that quasars and active galaxies can be powered only by gravitational energy, released at the center of the cosmic object. Gravitational energy derives from the attractive force between any two masses and is proportional to the product of the two masses and inversely proportional to their distance apart.

The first quantitative understanding of gravity was accomplished by Isaac Newton in the mid-seventeenth century; the latest theory of gravity, although still incomplete, was developed by Albert Einstein in 1915. Gravitational energy is tapped in the following way: When a small mass is allowed to fall toward a big mass—for example, a bowling ball dropped to the Earth—the smaller mass picks up speed as it falls. This increasing speed represents the release of gravitational energy, and it can be converted into other forms of energy, such as heat or radiation. If the big mass is highly concentrated, the small mass falling toward it will reach an enormous speed. A black hole is such a concentrated form of matter. Any mass falling toward a black hole reaches the speed of light before entering the hole. Such high speeds translate to efficiencies of 10 percent and higher at converting mass into energy, sufficient to account for the high energy produced in active galaxies and quasars.

According to current theoretical ideas, a massive black hole, with a mass of a million to a billion times that of our sun, inhabits the middle of each active galaxy or quasar. Surrounding gas and stars fall inward under the gravitational grip of the central black hole. As gas plunges toward the black hole, it releases its gravitational energy, which is then channeled into high-speed particles and radiation. Since the bulk of the energy release occurs just outside the black hole, and a massive black hole would occupy only a tiny volume at the center of a galaxy, the massive-black-hole hypothesis naturally explains the highly centralized nature of the radiation from active galaxies and quasars.

The concept of a black hole was first advanced in 1783 by

John Michell, the rector of Thornhill in Yorkshire, England. In 1796 the concept was rediscovered by Pierre-Simon Laplace. Here's the idea: For any body of a given mass and size, there is a critical speed, called the "escape speed," that is needed for a smaller body to escape the gravity of the larger. For example, the escape speed for the Earth is seven miles per second. Any mass thrown upward from the Earth with a speed less than seven miles per second will not be able to escape the gravity of the Earth; it will reach a maximum height and fall back to Earth. Any mass thrown upward with a speed greater than seven miles per second will escape the gravitational pull of the Earth and will travel outward into space, never to return. Now imagine taking the Earth and putting it into a giant vise, reducing its size by four times. The strength of gravity is stronger now because an object on the surface of the Earth is closer to the center of attraction. Now the escape speed has doubled, to fourteen miles per second. Continue squeezing the Earth to a smaller and smaller size, keeping its mass the same. With every compression of the Earth's size by four, the escape speed doubles. Eventually, when the size of the Earth is about two thirds of an inch, the escape speed has reached 186,000 miles per second, the speed of light. Now not even a light ray can escape the gravitational pull of the Earth. The Earth would appear black from the outside. It would be a black hole. Michell and Laplace, who first conducted this imaginary experiment, were familiar with Newton's theory of gravity and also knew the speed of light, which had first been measured in the seventeenth century.

Of course, there is no giant vise to squeeze the Earth down to a fraction of an inch. But, as we have seen, stars can squeeze themselves down to a very small size after they have exhausted their nuclear fuel and can no longer support their weight. In 1916, using Einstein's new theory of gravity, the German physicist Karl Schwarzschild refined the calculations of Michell and Laplace. The critical radius a mass must be compressed to in order to form a black hole is proportional to the amount of the mass and is about two miles for a mass equal to that of our sun. The critical radius is now called the "Schwarzschild radius."

In 1939 the American theoretical physicist Robert Oppenheimer and his student Hartland Snyder showed that a massive, burned-out star—such as a neutron star of several solar masses or more—would indeed collapse to form a black hole. However, even after Oppenheimer's calculations, most scientists still believed that black holes existed only on paper. Then, in 1965, astronomers discovered an object likely to be a black hole: Cygnus X-1, about 7,000 light-years from Earth. In the following decade, evidence emerged that Cygnus X-1 (named so because nearby gas emits X-rays and is located in the constellation Cygnus) is too massive to be a neutron star and too small to be a normal star. The only plausible alternative is that Cygnus X-1 is a black hole. Its mass is estimated at about ten times that of our sun. Since the mid-1970s, several other good black hole candidates have been identified.

As early as 1964, before any observational evidence for black holes, the Soviet physicist Yakov B. Zel'dovich and, independently, the American physicist Edwin Salpeter postulated that gas falling onto massive black holes might provide the power source for active galaxies and other energetic cosmic objects. A key to understanding quasars and active galaxies is the mechanism of feeding gas to the central black hole. Is it constant or intermittent? What triggers it? Possible sources of gas include ambient gas in the central regions of the galaxy, the gravitational shredding of hapless stars that wander too close to the black hole, the disintegration of stars by collisions with each other, and the agitation of one galaxy by a close encounter with another. For the more luminous active galaxies and the less luminous quasars, gas must be fed to the central black hole at the rate of about one sun's worth of mass per year. Some black holes may be spinning and release energy by slowing down, rather than by drawing directly on the gravitational energy of surrounding gas. However, such a process still requires surrounding material, to provide friction against the rotating hole. Isolated black holes, no matter how massive, produce very little energy. Thus, an understanding of the environment of the central black hole may be crucial to understanding why some galaxies are highly energetic and others are not.

How can we test the hypothesis of massive black holes? A massive black hole, even as massive as a billion times the mass of our sun, would have a diameter of only four billion miles. At the distance of the *nearest* big galaxy to ours, 2 million light-years away, such a black hole would have an angular size of only two hundred-millionths of a degree, far too small to be seen as a black void by any telescope in the near future. However, a massive black hole might reveal itself by the way it affects the motions and positions of surrounding stars. Trapped by the gravity of the hole, surrounding stars would huddle together more closely and would hurtle through space more rapidly than if no black hole were present. These effects might be noticeable out to a distance of several light-years from the black hole. The Hubble Space Telescope, and possibly the new eight-meter and ten-meter visible-light telescopes planned for the 1990s, may be able to discern such a telltale effect in nearby galaxies. Hints of a central concentration of stars have already been found in a couple of nearby galaxies, including the Andromeda galaxy. In galaxies beyond about 10 million light-years, however, the affected central region of the galaxy would be too small in angular size to see even through the new telescopes. At the present time, radio telescopes have a thousand times better angular resolution than visible-light telescopes and thus can see much finer details, although most stars do not emit much radiation at radio wavelengths. Radio telescopes can peer down to the central few light-years of active galaxies and quasars even billions of light-years away. If a massive black hole leaves radio fingerprints, they should be seen by high-angular-resolution radio telescopes, such as the Very Long Baseline Array planned for the 1990s.

Black holes might also be indirectly identified by the high-energy emission of the surrounding gas. Theoretical astrophysicists believe that gas near a black hole would orbit it in a flattened disk, similar to a protoplanetary disk, only hotter and more massive. The gas in the disk may heat up to temperatures between 1 million and 1 billion degrees and would consequently radiate mainly X-rays. Such X-ray emission will be studied by the Advanced X-ray Astrophysics Facility. Another characteristic feature of such high energies could be the production of electrons and their antiparticles, positrons. (Every subatomic particle has a twin, called its antiparticle, identical in most respects but with some properties just the opposite. The positron, for example, is identical to the electron except that it has an opposite electric charge.) Once produced, particles and antiparticles annihilate, producing gamma rays. The Gamma Ray Observatory and other proposed gamma-ray detectors in space will search for such gamma rays.

It is conceivable that the X-rays produced by active galaxies and quasars could account for the mysterious "X-ray background" that has puzzled astronomers for twenty-five years. When astronomers look out into space with X-ray detectors, they see a dense fog of cosmic X-rays in every direction. There is no consensus on the origin of the X-ray background. Proposed explanations have included the possibility of a hot gas filling the universe. Some scientists have suggested that the fog of cosmic X-rays may, in fact, arise from the X-rays emitted by active galaxies and quasars, too far away and too numerous to identify as individual sources of radiation. The resolution of this problem could bear upon the number of active galaxies and quasars in space and their evolution in time. On board the Advanced X-ray Astrophysics Facility will be detectors with the sensitivity and angular resolution needed to test for the first time the individual-source hypothesis for the X-ray background.

The dramatic behavior of active galaxies and quasars raises other questions. Some of the observed gaseous "jets" emanating at great speed from these objects are extremely narrow and well collimated. What produces and controls such columns of matter? Recently, some progress has been made on these questions. The jets radiate X-rays, visible light, and radio waves. Images made with large visible-light telescopes, radio telescopes, and the Einstein X-ray Observatory of the 1980s have revealed exquisite details of the structures and blobs in the gaseous jets. In particular, radio observations with very high angular resolution show that new concentrations of gas enter into the jet every year or so. As each gaseous blob moves out, it dims. We do not know whether the gaseous blobs closely follow each other, like beads on a string, or move independently, without constraint. Further detailed measurements of the radio emission show evidence for strong magnetic fields aligned along the direction of the jet. Indeed, some theorists propose that magnetic forces circle and squeeze the flowing gas, keeping it confined to a narrow column. The upcoming Very Long Baseline Array—a group of ten radio telescopes located throughout the United States, from Hawaii to the Caribbean, and acting in concert as one giant telescope—will allow a systematic study of the detailed structures and magnetic fields in

Fig. 15a. Photograph in visible light of the galaxy M87 *(also known as* Virgo A), *about fifty million light-years away.* Note *the faint blobby streak emanating from the center of the galaxy.*

the jets of a large number of active galaxies. Even better angular resolution will come from the Orbiting Very Long Baseline Interferometers (OVLBI), a group of state-of-the-art radio detectors orbited in space and linked to ground instruments to produce a telescope effectively larger than the entire Earth. The first OVLBI satellites will be launched by Japan and the former Soviet Union in the coming decade and have been developed with the help of American scientists.

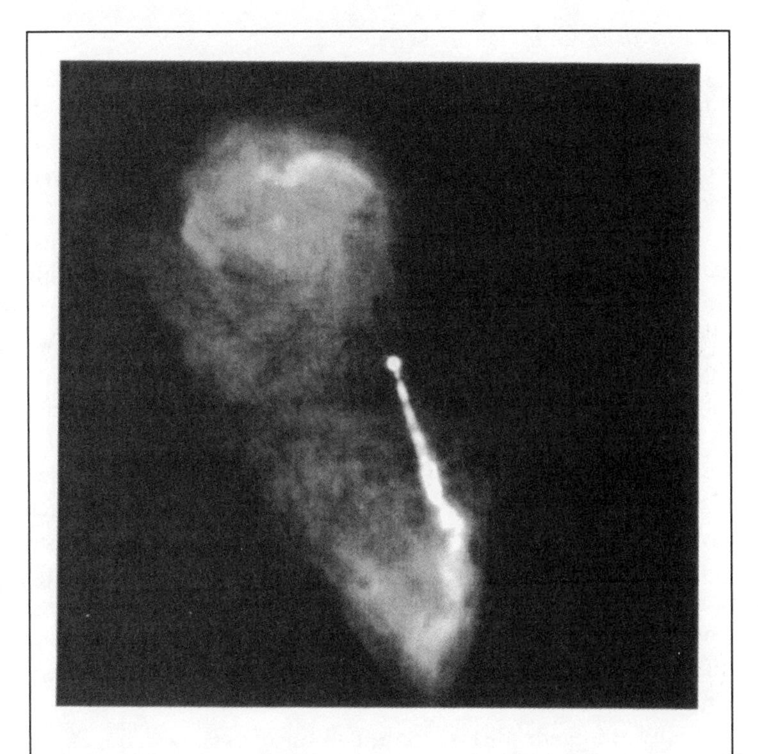

Fig. 15b. Photograph in radio waves of M87. The faint streak in the previous figure is now clearly seen as a well collimated gaseous jet. The jet is about 5,000 light-years long.

Continued theoretical work in the coming decade will also be crucial in understanding galactic jets. In the 1980s, some of the features of jets were reproduced in large computer simulations. In such simulations, the scientist programs the computer with the basic laws of physics describing how gas, radiation, and magnetic fields behave, sets up some initial configuration of matter and radiation, and then lets the computer calculate how the system evolves in time. Comparison to observation then guides refinements of the theory.

The Birth of Galaxies

Light received from greater distance was emitted earlier in time. If we peer out far enough in space, we should see back to the epoch of galaxy formation. Little is known about infant galaxies and even less about galaxies being born.

In 1928 the British astronomer James Jeans proposed the first quantitative theory for the formation of galaxies. According to this theory, the mass of the universe would have originally been scattered about very evenly, like sand on a flat desert. Here and there, however, tiny mounds would have poked up from the desert floor, containing a small concentration of extra mass. Jeans took the existence of the primordial mounds as a given, assuming that they would be explained by a later theory. In each mound, where there was extra mass, gravity would be stronger than average and so would attract the surrounding mass. This attraction would produce an even larger concentration of mass; the mound would grow in size, further increasing its gravitational pull on surrounding mass, and the process would accelerate. Eventually, the tiny mounds would grow into big hills. These big hills were the galaxies. There would be little mass left between galaxies.

Just as for individual stars, the collapse of a giant gaseous cloud to form a galaxy would occur only for mass accumulations sufficiently large that the inward pull of gravity could overcome the outward push of pressure forces. Also as for single stars, the formation of galaxies would be affected by radiation and rotation of the entire mass. In any case, small initial mounds would condense first, forming single galaxies; larger mounds would condense later, producing groups and clusters of galaxies; and so on, with a hierarchy of larger and larger structures. Indeed this model of galaxy formation is called the gravitational hierarchy model. The gravitational hierarchy model was substantially developed in the 1960s and 1970s by theorist James Peebles of Princeton University.

In the late 1960s and early 1970s, Soviet astrophysicists Yakov B. Zel'dovich, A. G. Doroshkevich, and their colleagues proposed a competing model of galaxy formation called the pancake model. According to this model, the force of radiation would have leveled any small initial mounds of mass. The flatness of the primeval cosmic desert would have been broken only by rather broad mounds, containing at least 1,000 times the mass of a single galaxy, even though such mounds would have poked up only slightly above the desert floor, as in the case for the gravitational hierarchy model. As these larger concentrations of mass radiated their heat and cooled and began collapsing, they inevitably would have collapsed fastest along one axis, thus producing a thin "pancake" of gas. The massive pancake would then have fragmented into many pieces, with each piece becoming an individual galaxy. According to this picture of galaxy formation, one might expect that over a region corresponding to each parent pancake, galaxies would be distributed in thin sheets.

Theorists argue over which, if either, of these two models is correct. The question hinges partly on what caused the initial mounds in the early universe, whether those initial mounds were connected over large distances, and what types of subatomic particles were present at that time. Other factors could be involved. In both the gravitational hierarchy model and the pancake model, gravity plays the dominant role in forming galaxies. Some maverick astronomers, notably Jeremiah Ostriker of Princeton and Lennox Cowie of the University of Hawaii, have suggested that, instead, gas pressure forces might be the prime movers, in which case the initial mounds in the early universe would be less relevant. It seems plausible that, at the very least, gas pressure forces and radiation played some role in determining galactic sizes and masses, which do not vary much from one galaxy to the next. In the last decade, Ostriker and his collaborators have done detailed computer simulations, including gas pressure forces, to explore the consequences of the theoretical models. It seems that gravity, gas pressure, and the initial rotation of the pregalactic gas cloud are all important in forming a galaxy.

On the observational side, astronomers are looking for a variety of clues. Are existing galaxies located in randomly shaped groupings of ever-larger size, as predicted by the gravitational hierarchy model, or are they distributed in sheets of one basic size, as predicted by the pancake model? Indeed, the pancake model suggests the existence of primordial pancakes of gas, each containing about 1,000 times the mass of one galaxy. These should be seen at distances of about 10 billion light-years. Such giant structures, the possible birthplace of galaxies, are composed mostly of hydrogen gas and emit distinctive radio waves at a wavelength of twenty-one centimeters, or about eight inches. The radiation will be searched for with the Very Large Array, the new Green Bank Telescope, and the upgraded Arecibo Telescope.

Even more fundamental issues are at stake. In both the gravitational hierarchy model and the pancake model, strong concentrations of mass result from the force of gravity acting on initially tiny inhomogeneities. Although tiny, initial inhomogeneities are required for any condensation to occur. How can we find them? The answer lies most likely in the cosmic background radiation. In 1965 Arno Penzias and Robert Wilson, working at Bell Laboratories in New Jersey, discovered a bath of radio waves called the cosmic background radiation, filling all space. It is believed that this radiation has been traveling freely through space since the universe was only about 300,000 years old. Before then, the radiation was rapidly bouncing off the subatomic electrons scattered through the cosmos, like balls bouncing off the bumpers of a pinball machine. Imprinted on the cosmic background radiation should be a record of the distribution of cosmic matter at the last "bounce," when the universe was 300,000 years old, well before the epoch of galaxy formation. If at that time the matter of the universe was very smoothly distributed, like bumpers evenly spaced and very close together, then the cosmic background radiation should have bounced around equally in all directions and emerged uniformly. By contrast, any unevenness in the distribution of matter would have caused unevenness in the cosmic background radiation. That ancient unevenness would show up today as a variation in the radiation intensity as our radio telescopes point in different directions—and indeed all theories of the formation of galaxies demand the existence of such variations.

After decades of searching, the variations have been finally observed, at the level of one part in one hundred thousand. This extremely important discovery was made in April 1992 by the Cosmic Background Explorer (COBE), a satellite launched in 1989 with the express purpose of measuring the cosmic background radiation to high precision. These results will serve as an important constraint on all theories of galaxy formation and large-scale cosmic structure. First, they tell us how big were the tiny condensations of matter that gravity would have to work on in order to eventually produce galaxies and groups of galaxies. Second, they challenge us to explain the size of the initial condensations themselves. At the present time, no theory of galaxy formation and of the early universe provides a compelling explanation of COBE's results. Over the next few years, several new candidate theories will be worked out in enough detail to make a firm comparison to the observational results. This is surely an exciting time for cosmological theory.

Finally, important hints to galaxy formation lie in the gas between galaxies, called the intergalactic medium. As in the interstellar medium within individual galaxies, this gas is enriched with the various chemical elements manufactured within stars. And the intergalactic medium at large distances—that is, at early times—is the material out of which galaxies formed. A primary tool for analyzing the intergalactic medium has been the study of radiation from quasars. As this radiation travels from there to here, it passes through the intergalactic medium, and some of it is absorbed. The particular wavelengths absorbed indicate the particular chemical makeup of the intervening gas. How does the chemical makeup of the intergalactic gas change in time? In other words, how does it change with distance from Earth? Can this gas be used to date the epoch when galaxies first formed? Analysis of the intergalactic medium at large distances requires both high sensitivity and high spectral resolution. The needed abilities are beyond current ground-based 4-meter telescopes but within reach of the much larger telescopes of the coming decade.

Garth Illingworth was born on March 17, 1947, in Perth, Australia. He received a B.Sc. degree in physics from the University of Western Australia in 1968 and a Ph.D. in astrophysics from Australian National University in 1973. After postdoctoral fellowships, Illingworth was an astronomer at Kitt Peak National Observatory in Tucson, Arizona, from 1978 to 1984. From 1984 to 1987, Illingworth was deputy director of the Space Telescope Science Institute in Baltimore, during which time he was also a research professor at Johns Hopkins University. In 1988 he became professor of astronomy at the University of California at Santa Cruz. Illingworth's research has centered on the observations of star clusters and galaxies, with the goal of understanding how galaxies formed and evolved. Illingworth has worked on the Keck telescope, and he has initiated an effort to create a large space telescope to succeed the Hubble Space Telescope.

Says Illingworth, "One of the most delightful and rewarding aspects of being an astronomer is that dedicated individuals and small teams can make great progress on the outstanding problems of astrophysics. Of these problems, one of the most difficult is understanding when galaxies like our own Milky Way formed, how they formed, and how they have changed with time. We know that galaxies condensed in the youthful days of the universe from the tenuous gas and dark matter that permeated space, to become the concentrated, massive objects we now see. But we do not know how or exactly when this happened. I expect that the powerful new telescopes and computers of the 1990s will help us push back this frontier and open up the young universe to our eyes."

The Life History of the Universe

The Big Bang Model

As we look deeper and deeper into space, will we come to an edge of space, or a beginning of time? If the universe had a beginning, how did it begin? Will it have an end? These are questions in cosmology, the branch of astronomy concerned with the structure and evolution of the universe as a whole.

Every culture has invented a cosmology. Aristotle's universe had an edge of space, an outermost crystalline sphere upon which were fastened the stars. But his cosmos had no beginning or end of time. Futhermore, it was static. In *On the Heavens,* Aristotle wrote that the "primary body of all is eternal, suffering neither growth nor diminution, but is ageless, unalterable and impassive." He conjectured that the heavens, being divine and immortal, must be constructed of this primary body, the "aether." The Judeo-Christian worldview did away with eternity, but maintained the idea of a cosmos without change. In this tradition, the universe was created from nothing by God and has remained much the same ever since. Copernicus, who in 1543 demoted the Earth from the center of the cosmos to a mere planet in orbit about the sun, changed many things, but not the Aristotelian belief in a universe both spatially finite and static in time.

In 1576 Englishman Thomas Digges became the first Copernican to pry the stars off their crystalline spheres and spread them throughout an infinite space. Digges, born in 1546 and trained in mathematics, published an English translation of portions of Copernicus's great work, *De Revolutionibus.* In his translation, titled *A Perfit Description of the Caelestiall Orbes,* Digges added a section about his own ideas regarding the distribution of stars, especially his notion that space was infinite and that stars were scattered through this infinite space. (Digges was also interested in more practical matters. He made the earliest serious study of the trajectory of artillery shells and became the father of ballistics in England.)

After Digges the universe was considered without limit in space. However, people still viewed the universe as unchanging in time. The great physicist Isaac Newton argued the same view a century later. Individual planets moved in the sky, to be sure, but the universe as a whole was thought to look pretty much the same from one eon to the next.

Newton correctly understood that gravity was the dominant force for describing the cosmos at large. Furthermore, with his theory of gravity, Newton was the first person capable of making a quantitative model of the universe as a whole. However, no such model was put forth until that of Albert Einstein, in 1917. Einstein, like Newton, had recently developed a new theory of gravity, called general relativity. Like Newton, Einstein understood that gravity was the principal force to be reckoned with in cosmology.

General relativity is a highly mathematical and complex theory of how gravity is generated by matter and energy and how that matter and energy, in turn, respond to gravity. In order to solve the difficult equations of his theory for the universe as a whole, Einstein made two simplifying assumptions: the universe does not change in time, and the matter of the universe is evenly scattered through space. Although there was no observational evidence for either of Einstein's starting assumptions, he had faith that they might be close enough to the truth to give satisfactory results. Einstein's resulting "cosmological model" was static and homogeneous.

It soon became clear that very different cosmological models were possible. In 1922 Alexander Friedmann proposed a cosmological model for a *nonstatic* universe, a universe that changed in time. Friedmann, a Russian mathematician and meteorologist, began with Einstein's theory of gravity and accepted his assumption of homogeneity but questioned the assumption of stasis. As the Dutch astronomer Wilhelm de Sitter had pointed out, our view of the universe—even through a large telescope—provides only a snapshot, offering little information about long-term behavior. Friedmann sought and found a different solution to the equations of general relativity, a solution in which the universe began in a state of extremely high density and then expanded in time.

In Friedmann's cosmological model, the matter of the universe would became more and more dispersed as it expanded from its primal explosion. This cosmological model came to be called the "big bang model." In a reply in 1923 to Friedmann's evolving model, Einstein acknowledged the mathematical validity of Friedmann's calculations but doubted their applicability to the real universe. It is not uncommon in theoretical physics to find more than one solution to a set of equations, depending on the beginning premises, and Einstein remained convinced, like Aristotle and Copernicus and Newton before him, that the real universe was static. However, the starting assumptions of neither Einstein nor Friedmann could be empirically tested. At the time, there was practically no experimental evidence one way or the other. Einstein and Friedmann had worked out pencil-and-paper universes.

The situation changed dramatically in 1929. In that year the American astronomer Edwin Hubble discovered, through the eyepiece of a telescope, that the universe is expanding. The galaxies are sailing away from each other.

Hubble did not actually see the galaxies drift across the field of view of his telescope; such motion would take many millions of years to see directly. Rather, Hubble inferred the galaxies' motions by their Doppler shifts: their colors were shifted toward the red end of the spectrum, a shift that indicated outward motion and that came to be called the "redshift." All the galaxies were moving outward from the Milky Way. Actually, the redshift of many cosmic nebulae had been measured in the early 1900s by Vesto Slipher at the Lowell Observatory in Arizona. What Hubble added to Slipher's work was a determination, using Cepheid stars, of the *distances* to the receding galaxies. Hubble discovered that the distance to each galaxy was proportional to its recessional speed. In other words, a galaxy twice as far from us as another galaxy was moving outward twice as fast. This result was just as expected for a universe stretching uniformly in all directions.

On the one hand, Hubble's observation clearly favored Friedmann's nonstatic model over Einstein's static one. On the other, Hubble's observation apparently confirmed one key assumption made by both scientists: the universe is approximately homogeneous. Only if the universe is homogeneous will the galaxies move outward with a speed proportional to distance. Furthermore, a homogeneous universe means that no location is different from any other. The universe is expanding, but there is no center of the expansion, just as there is no center of expansion on the *surface* of an expanding balloon. If dots are painted on such a balloon, each dot representing a galaxy, then from the vantage of *any* dot all the other dots are moving away from it. No dot is the center.

If the outward velocities of galaxies are proportional to their distances, then the ratio of velocity to distance is the same for any galaxy. This ratio, called the Hubble constant, measures the current rate of expansion of the universe. According to the best measurements, the current rate of expansion of the universe is such that the universe will double its size in approximately 10 billion years. More precisely, the distance between any two distant galaxies will double in approximately 10 billion years.

As time goes on, the galaxies move farther away from one another. It follows necessarily that in the past, they must have been closer together. If we imagine a movie of the universe played backwards in time, the galaxies would crowd closer and closer together, until some definite moment in the past when all the matter of the universe was compressed together in a state of almost infinite density.

Astonomers can estimate when this point in time occurred: about 10 or 20 billion years ago. That moment is called "the big bang." What happened before the big bang is a subject of intense speculation among theoretical physicists.

When astronomers first estimated the age of the universe, in the 1930s, they compared that age with the age of the Earth. As mentioned earlier, radioactive dating of uranium ore, beginning around 1910, suggests that the Earth is about 4.5 billion years old. Almost all theories of the formation of the sun and Earth require that the Earth be somewhere between 10 percent and 90 percent the age of the universe. In other words, from dating the rocks in the ground, scientists predict that the universe is between 5.5 billion and 50 billion years old. And from the motions of galaxies, scientists predict that the universe is between 10 and 20 billion years old. These are two very different types of measurements. Their agreement is powerful evidence in favor of the big bang model. However, it is important to remember that cosmology, of all branches of astronomy and indeed of all science, requires the most extreme extrapolations in space and in time. The big bang model, although widely held, rests on a rather small number of observational tests.

The two principal tests, after the test against the age of the Earth, are the chemical makeup of the cosmos—approximately 74 percent hydrogen, 24 percent helium, and 2 percent heavier elements—and the smooth bath of radio waves from space, called the cosmic background radiation.

According to the big bang model, both the bulk chemical makeup of the cosmos and the cosmic background radiation were created long ago, when the universe was very different from what it is today. If we again play our movie of cosmic evolution backward in time, the universe contracts, the galaxies move closer and closer together, and eventually the galaxies and stars lose their individual identity. The matter of the universe begins to resemble a gas. As the universe continues to contract, growing denser and denser, the cosmic gas becomes hotter and hotter. When the temperature has reached about 10,000 degrees centigrade, the electrons escape from their atoms. At a still higher temperature, the atomic nuclei disintegrate into protons and neutrons. As the origin of the universe, the big bang, gets nearer and nearer, the temperature continues to mount. At a temperature of 10 trillion degrees, each proton and neutron disintegrates into three elementary particles called quarks.

Now imagine going forward in time, from the beginning. About 0.00001 of a second after the big bang, the quarks combined into protons and neutrons. The nucleus of hydrogen, the lightest chemical element, is made of a single proton. However, no other chemical elements could exist at this time. All other elements consist of a fusion of two or more subatomic particles, which could not hold together under the intense heat of the infant universe. As the universe expanded, it cooled. When the universe was a few minutes old, its temperature had dropped to a billion degrees, the critical temperature at which neutrons and protons could begin sticking together via the attractive nuclear forces between them. This is when deuterium, helium, and lithium would have been formed, as calculated by theorists in the 1960s and 1970s. The first such calculations were done in 1964 by Fred Hoyle and Roger Tayler of Cambridge University and by Yakov B. Zel'dovich of the Institute for Cosmic Research in Moscow. Further calculations were done by James Peebles of Princeton, Robert Wagoner and collaborators at the California Institute of Technology, and David Schramm and collaborators at the University of Chicago. Remarkably, these theoretical calculations are in accord with the observed abundances of hydrogen, helium, lithium, and deuterium. (All heavier elements, such as carbon, oxygen, and iron, were manufactured much later, by stars.) This agreement is another piece of evidence supporting the big bang model.

The hot infant universe would have also produced the cosmic background radiation. According to theoretical predictions first made in 1948 by Ralph Alpher, George Gamow, and Robert Herman of George Washington University and then repeated independently in 1965 by Robert Dicke and James Peebles of Princeton, a special kind of radiation called blackbody radiation would have been produced throughout space when the universe was younger than a few seconds old. Blackbody radiation can be characterized by a single parameter, which corresponds to the temperature of the radiation. In theory, blackbody radiation should have been produced uniformly through space in the early universe and would have continued bouncing off subatomic particles until the universe was about 300,000 years old, when electrons and atomic nuclei combined to make atoms. After that point, the radiation would have simply streamed through space, oblivious to matter. As the universe expanded, the radiation would have continuously increased its wavelength until the present time, when it should have a wavelength corresponding to radio waves and a temperature of about three degrees above absolute zero. As mentioned in the last chapter, this radiation was discovered accidentally in 1965. The recent Cosmic Background Explorer has confirmed that the cosmic background radiation has precisely the properties predicted by the big bang theory, thus providing one more crucial piece of evidence in support of the theory.

The Large-Scale Structure of the Universe

The big bang model assumes that the universe is homogeneous. Clearly, this assumption cannot be exactly true. A homogeneous universe would not have the "lumps" that are stars and galaxies and clusters of galaxies. It is possible, however, that such lumps are significant only when viewing the universe on a small scale. It is possible that space might appear smooth when averaged over a sufficiently large volume, just as a beach appears smooth when looked at from a distance of ten feet or more, even though it appears grainy when looked at from closer range. The real question is, how far back from the beach must we look? If there is no scale on which the universe looks smooth, then the big bang model could be in serious trouble.

Evidence for rather large-scale inhomogeneities has come in recent years, with the discoveries of chains of galaxies, sheets of galaxies, and giant voids with very few galaxies at all. Such large groupings of galaxies are called "structures." A completely homogeneous universe would have no structures at all. In 1989 Margaret Geller and John Huchra of the Harvard-Smithsonian Center for Astrophysics found evidence for the largest structure so far known, a "wall" of galaxies stretching at least 500 million light-years in length. Geller and Huchra have, in effect, constructed three-dimensional maps of large groups of galaxies. More precisely, these scientists and others have recorded the two-dimensional position on the sky and the redshift for each of many galaxies. Remember that the redshift is a quantitative measure of outward speed. If we assume that the universe is approximately homogeneous and uniformly

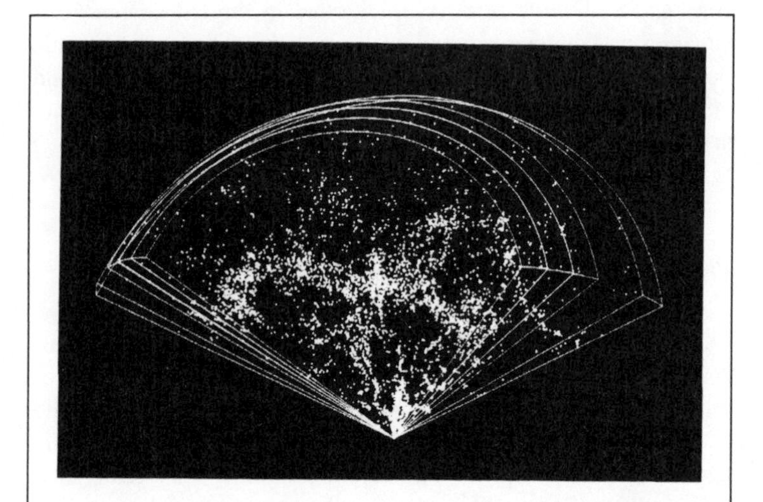

Fig. 16. Redshift survey by Margaret Geller and John Huchra of the Center for Astrophysics, completed in 1989. Each dot is a galaxy; there are 3,962 galaxies here, in several touching wedges in space. The radial direction directly measure recessional speed, which corresponds to radial distance in a homogenous universe. The farthest galaxy shown is about 500 million light-years away. The nearly continuous horizontal bunch of galaxies stretching across the diagram has been called the "Great Wall" and appears to be the largest coherent structure of galaxies yet observed.

expanding—so that we can apply Hubble's law of proportionality between radial distance and outward speed—the redshift of a galaxy translates into an approximate radial distance. In this way, all three dimensions are known, and a 3-D map can be made. (Roughly, the approximate distance to a galaxy is 10 billion light-years times the fractional increase in wavelength of detected radiation. For example, a wavelength increase of 10 percent corresponds to a distance of 1 billion light-years.)

It is not yet known whether the new cosmic structures found in a few selected regions of space are typical. So far, however, structures have usually been found at every possible scale. A survey that covers 50 million light-years usually finds some chain or disk or absence of galaxies extending roughly 50 million light-years in size; a survey covering a 500-million-light-year region finds structures of 500 million light-years, and so on. Some astronomers speculate that cosmic structures of all sizes exist.

Challenging the viewpoint of a hierarchy of structures of ever-increasing size are recent surveys by T. J. Broadhurst and R. S. Ellis of the University of Durham in England, David C. Koo of the Lick Observatory in California, Richard Kron of the University of Chicago, and Alex S. Szalay of Johns Hopkins University. These surveys, called "pencil beam" surveys, measure the redshifts of galaxies only along one outward line through space. However, the pencil beam goes out very far, to a distance of several billion light-years. The pencil beam surveys suggest that galaxies bunch together roughly every 400 million light-years, but not on larger scales. If so, the largest coherent cosmic structures would be about 400 million light-years in size, about the length of the "Great Wall," and the universe would appear homogeneous when averaged over much larger distances. Much more observational work is needed to confirm or refute these suggestions.

More complete galaxy surveys are now in progress that extend out to a few billion light-years. In addition, surveys with many more galaxies are being planned. The largest redshift surveys to date include only several thousand galaxies. In the coming decade a team of astronomers from Princeton University, the Institute for Advanced Study, and the University of Chicago will initiate a completely computerized survey of the sky to measure the positions and redshifts of about *one million* galaxies. The team will use a wide-angle, one-hundred-inch telescope designed by James Gunn of Princeton, to be installed in the Sacramento Mountains of New Mexico. Such a survey can be accomplished with a moderate-sized visible-light telescope, if dedicated to the job and augmented with new fiber optics that can record the individual emissions of many galaxies at once.

Similarly large redshift surveys are planned for radio telescopes, particularly the upgraded Arecibo Telescope and the Green Bank Telescope. As mentioned earlier, atoms of hydrogen gas in galaxies emit radio waves with a well-defined wavelength, and the increase in wavelength (redshift) of such emission will be accurately measured for large samples of galaxies.

Whatever inhomogeneous groupings of galaxies are found, the high degree of uniformity of the cosmic background radiation means that the universe *ought* to be homogeneous on scales of 10 billion light-years. Such homogeneity would be contradicted by cosmic structures with sizes of a few billion light-years, only several times larger than those now being mapped. A confrontation is coming. In the next decade, astronomers will almost certainly find such large structures, if they exist.

Inhomogeneities can be studied in other ways. In a uniformly expanding universe, the outward velocity of any galaxy is strictly proportional to its distance. But this relationship fails when the material of the universe is not homogeneous. In that case, the motions of galaxies are altered by the irregularity of the gravity they feel. Such altered motions are called "peculiar velocities" and can be used as diagnostics of the inhomogeneities of mass. In 1987 astronomers David Burstein, Roger Davies, Alan Dressler, Sandra Faber, Donald Lynden-Bell, R. J. Terlevich, and Gary Wegner discovered that a group of galaxies about 200 million light-years away has substantial peculiar velocities, as if all attracted by some large mass. That mass has been called the Great Attractor. Peculiar velocities have been found for other groups of galaxies as well.

It is not clear that the observed inhomogeneities of matter can fully account for the observed peculiar velocities. If not, then either there must be a substantial amount of *lumpy* dark matter between galaxies, invisibly deflecting them in their travels, or additional forces besides gravity must be at work. Neither of these two possibilities would be welcome to astronomers. In the near future, astronomers hope to measure the peculiar velocities of about 15,000 galaxies out to a distance of about 300 million light-years.

Independent measurements of cosmic distances are critical for these studies and for all studies of large-scale structures. It cannot be assumed that distance is proportional to redshift—such an assumption is equivalent to the assumption of homogeneity, which is precisely what is being tested. In recent years, progress has been made in developing new ways to measure vast distances in space, starting in our own galaxy. The precision of radio detectors at measuring positions and changes of positions of moving objects has permitted accurate distance determinations to the center of our galaxy and to nearby galaxies. Astronomers hope eventually to place in space a special type of telescope, called an optical interferometer, that will be able to measure angular changes in position as tiny as *three billionths of a degree*, the width of a penny at the distance of the moon. With this extraordinary precision, the distance to every Cepheid star in our galaxy can be accurately determined by measuring how its apparent position shifts as the Earth orbits the sun. It is crucial that we have good distances to Cepheid stars.

Unfortunately, Cepheid stars cannot be used to determine distances to most galaxies, since individual stars cannot be seen beyond about 30 million light-years. However, other yardsticks have been discovered. For example, in the mid-1970s, using nearby galaxies of known distance and luminosity, Sandra Faber of the Lick Observatory at the University of California at Santa Cruz and Robert E. Jackson of the Space Telescope Science Institute in Baltimore and, independently, R. Brent Tully of the University of Hawaii and J. Richard Fisher of the National Radio Astronomy Observatory in Green Bank, West Virginia, found an empirical relationship between the luminosity of a galaxy and the speed of its stars. Once this relationship has been calibrated for nearby galaxies, it can be applied to farther galaxies, where only the stellar speeds can be measured (by the Doppler shift in colors). Then, just as for Cepheid stars, the distance is determined from the inferred luminosity and the observed apparent brightness. The trouble with this method is that looking to greater distances in space is equivalent to looking back in time. The nearby galaxies used to calibrate the luminosity–stellar speed relationship are considerably older than the distant galaxies to which the relationship is applied. For this reason, it is crucial to understand how the properties of a galaxy—such as its luminosity and the motions of its stars—evolve in time.

A promising new method for measuring cosmic distances, developed by John Tonry of the Massachusetts Institute of Technology, involves variations in the numbers of stars seen by a telescope as it sweeps across a galaxy. An analogy will explain this method. Consider counting marbles scattered haphazardly across the floor. Because the marbles are randomly spaced, the number of marbles on each one-square-foot patch of floor will vary from one part of the floor to another. The number of marbles on every ten-square-foot patch of floor will also vary, but the *fractional* variation will be less because more marbles are included in each patch. In a similar way, opinion polls of the public have smaller fractional error when more people are included in the poll. Now replace the marbles with stars. A telescope viewing a *nearby* galaxy sees a relatively small patch of the galaxy within its field of view. Thus, as it surveys different patches of the galaxy, the telescope will record a relatively large fractional variation in the number of stars, or intensity of light, from one patch to the next. By contrast, a telescope viewing a *distant* galaxy sees a relatively large patch of the galaxy within its field of view, leading to a smaller fractional variation in light intensity as it sweeps across the galaxy. Once the patch-to-patch variation in light intensity has been calibrated for a nearby galaxy of known distance, the method can be used to determine the distance to much farther galaxies.

On the theoretical side, astronomers are attempting to make sense of the observed positions and motions of galaxies by the use of large computer simulations. Such simulations involve 10,000 to 10 million particles, each representing a portion of a galaxy or a number of galaxies. The particles are placed at initial positions and then allowed to interact and move via their mutual gravity. The final positions on the computer screen, after one hundred hours of computer time, represent the positions and motions of galaxies after 10 billion years of real time.

By comparing computer simulations to the observed large-scale structure of the universe, scientists hope to test their assumptions about the initial conditions and forces at work in the cosmos. The current computer simulations, although ten times larger than those of a decade ago, still do not have enough particles for a decisive comparison between theory and observation. Within the coming decade, larger computers and new methods for using those computers should give more reliable answers. And the new calculations will have more brain as well as brawn. Additional physics will be programmed into the computers, and the resulting simulations will be more realistic and believable. As mentioned earlier, astronomers have long assumed that gravity is the dominant force in forming galaxies and larger structures in the universe. Just since 1989, computer simulations have begun to include nongravitational forces, such as gas pressure.

Whatever the outcome of the computer simulations, the ultimate theory of the distribution of matter in the universe must be consistent with the observed cosmic background radiation. How could matter have been sufficiently lumpy in the past to have produced the large chains and walls of galaxies we see today, and yet have left the cosmic background radiation with lumpiness of only a few parts per million? When cosmologists are asked to list the outstanding problems in their field, the large-scale structure comes out on top.

Dark Matter

In the late 1970s, astronomers realized that at least 90 percent of the mass in the universe is invisible. This invisible mass can be detected by its gravitational effects on the stars and galaxies that we see, but it emits no electromagnetic radiation of its own—no visible light, no radio waves, no infrared, no ultraviolet, no X-rays, no gamma rays. It is truly invisible. It is called dark matter. The existence of dark matter definitely complicates our understanding of the large-scale structure of the universe.

The possibility of dark matter was first suggested by Fritz Zwicky in 1933. Zwicky observed a group of galaxies orbiting one another and estimated the gravity needed to keep the cluster from flying apart. From the required gravitational pull and the size of the

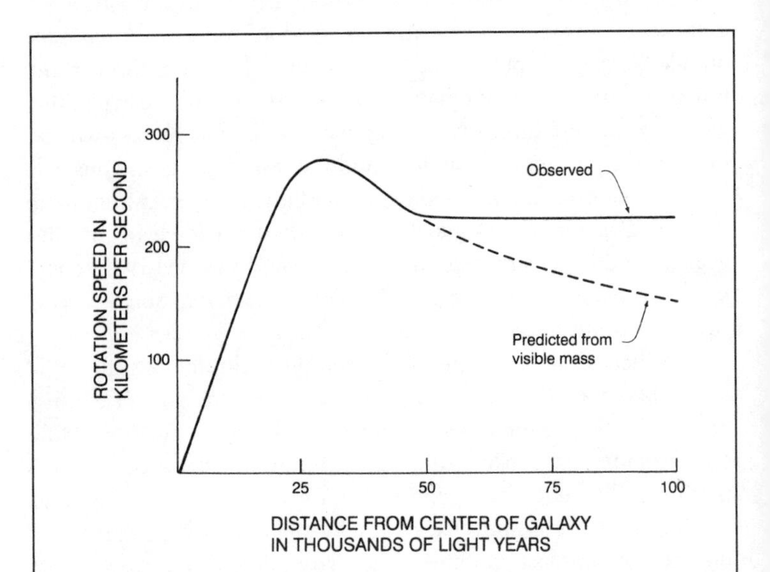

Fig. 17. Orbital speed of hydrogen molecules in the Andromeda galaxy at various distances from the center of the galaxy. The orbital speed at each position tells how much mass is interior to that position. If the only mass in the galaxy were the visible mass, the orbital speeds would continuously decrease at distances from the galactic center greater than about thirty thousand light-years. Instead, the orbital speeds level off after about fifty thousand light-years, indicating the presence of dark matter. Figure adapted from the observational data obtained by Vera Rubin, W. K. Ford, and Morton Roberts.

cluster, Zwicky could further calculate the mass contained within the cluster: about twenty times what could be generated by the visible stars and gas. Zwicky was regarded as something of an eccentric, and his conclusion was deemed preposterous. For more than forty years, most astronomers tried to ignore the possibility of dark matter. Then, in the 1970s, astronomers performed new observations of the orbital motions of stars and gas in individual galaxies.

Among the crucial observations that convinced the world of dark matter were those of Vera Rubin and collaborators at the Department of Terrestrial Magnetism of the Carnegie Institution of Washington. Rubin measured the speed of gas orbiting the center of the Andromeda and other galaxies, and from this speed she was able to infer the amount of mass needed to whirl the gas around.

At the time, in the 1970s, many astronomical images were still recorded with photographic plates, not with the digital detectors of today. Consequently, Rubin had to arrive at the observatory a full day before observations to begin preparing her photographic plates. Because she was measuring the faint light from the outermost parts of galaxies, her plates had to be supersensitive. In complete darkness, she had to cut the glass into two-inch squares, place the squares in an atmosphere of dry nitrogen, and bake them for hours to increase their sensitivity. A single dust mote would have spoiled the plates for the job they had to do. After the three-hour exposures, Rubin had to develop the plates, dry them, and stare at them through a microscope, measuring the locations of spectral lines to an accuracy of better than one ten-thousandth of an inch. Today, the entire procedure can be done automatically, with electronic detectors and computer analysis.

Throughout her career, Rubin has pioneered a number of projects that challenged conventional wisdom. Ironically, she prefers noncontroversial topics. At a meeting of the American Astronomical Society in Philadelphia, in 1950, Rubin presented a ten-minute paper giving evidence that a large group of galaxies were rotating together. This was, in fact, the first recognition of large-scale, peculiar velocities of galaxies, an extremely unpopular idea at the time. Rubin was then twenty-two years old and partly occupied with nursing her newborn at the meeting.

Vera Rubin was born in 1928 and raised in the Washington,

D.C., area. She remembers lying in bed at age ten and watching the stars move through the night, beyond her bedroom window. Rubin received a B.A. from Vassar in 1948, an M.A. from Cornell in 1951, and a Ph.D. from Georgetown University in 1954. Her doctoral work involved one of the first quantitative studies of the clustering of galaxies. Rubin recalls that when she first found evidence for dark matter, in 1978, no one doubted the data—the speed of gas at different distances from the center of galaxies—but many people wished that the data could be explained without dark matter.

James Peebles was born on April 25, 1935, in Winnipeg, Manitoba, and is a Canadian citizen, although he works in the United States. Peebles received a B.S. degree from the University of Manitoba in 1958 and a Ph.D. in physics from Princeton University in 1962. All his professional career has been spent at Princeton, where he is now the Albert Einstein Professor of Science. Peebles is a theorist. His main interest is cosmology. In 1965, in collaboration with Robert Dicke and others, Peebles predicted

the existence of the cosmic background radiation. The next year, Peebles did one of the first detailed calculations of the cosmic helium abundance expected from nuclear reactions in the early universe. In 1965 Peebles pioneered calculations of the gravitational clustering of matter in an expanding universe and has become a leading theoretician of the "gravitational hierarchy" models for large-scale structure in the universe. Many physicists first became interested in cosmology through Peebles's book Physical Cosmology *(1971).*

Says Peebles, "Research in science is guided by grand questions, but most of the work goes into details—small puzzles whose solution might advance our knowledge by a little. Here is my list of questions I expect we will be discussing in the 1990s: What is the nature of the dark matter around galaxies and in clusters of galaxies? If the mass of the universe is large enough to stop the cosmic expansion, what is this mass and where is it hidden? What is the geometry of space? Is it closed, flat, or open? How did galaxies form? Why do the galaxies tend to be found in sheets, in groups, and in clusters? What is the origin of 'ordinary things': particles and cosmic magnetic fields? Can we find some way to probe the physics of the very early universe, to test the inflationary universe model for the way our universe began? Can we find alternative models?"

What is dark matter? We know that it exists, but we have little idea what it is. Dark matter could be planets or extremely dim stars, strewn through space. Dark matter could be a vast sea of subatomic particles. Whatever it is, dark matter makes up the bulk of the universe.

And it is not just the unknown identity of dark matter that

causes concern. Its quantity and arrangement in space are also uncertain, foiling attempts to understand why the luminous mass is arranged as it is. Dark matter apparently exists on all scales. On the larger scales, a careful reconciliation of the peculiar velocities of galaxies with the *observed* inhomogeneities in luminous matter should reveal the presence of dark matter, which contributes to the peculiar velocities through its gravitational effects. Astronomers know that at least some of the dark matter is bunched and clustered around luminous matter. But is all of it so clustered? A smooth distribution of dark matter does not cause peculiar velocities of galaxies and so would be harder to find.

Dark matter will also be mapped in the coming decade by X-ray emission from hot gas. Very hot gas has been detected inside large clusters of galaxies and extends 5 to 10 million light-years out from the center of many clusters. So hot that it should boil away, the gas is evidently held by the gravity of invisible matter. From the precise distribution of the gas, astronomers can work back to infer the gravity confining it and the distribution of dark matter producing that gravity. In the coming decade, the German X-ray Roentgen Satellite, recently launched, the Japanese X-ray satellite Astro D, and the American Advanced X-ray Astrophysics Facility will make better maps of the hot gas in galaxy clusters. The latter two missions will also measure the temperature of that gas.

One new technique for measuring dark matter makes use of the "gravitational lens" phenomenon. Gravity attracts light rays as well as material bodies. Thus when light from a distant astronomical object, like a quasar, travels toward the Earth, that light should be deflected by any mass it passes on the way. The intervening mass can split and reshape the image of the quasar. By analyzing the distortions of quasar images, theoretical astronomers can infer the location of the intervening mass, even if that mass is invisible, like dark matter. Since 1979, when gravitational lenses were first discovered, about a dozen have been found. In the coming decade, the gravitational-lens phenomenon will be used as a powerful tool to map out and uncover the nature of dark matter. Such a program has already been started by Anthony Tyson of AT&T Bell Laboratories and others.

Some astronomers have proposed that the dark matter consists of large planets. Large planets are not quite invisible; they emit a low intensity of infrared radiation. The Space Infrared Telescope Facility should have the required sensitivity to find infrared-emitting planets in the far reaches of our own galaxy, where dark matter may be lurking.

Alternatively, dark matter could consist of freely roaming subatomic particles, rather than aggregates of particles such as planets. The possibilities have stirred the imaginations of particle theorists. Dozens of particles have been proposed, particles never before seen in the laboratory. These hypothetical particles have names like the axion and the photino and are predicted on the basis of new theories of subatomic physics. However, the properties of the new particles are uncertain. All that is known is that their effect on other matter should be extremely weak, since they have never been seen. If dark matter does indeed consist of these exotic particles, then it may be identified in the laboratory rather than in space. Within the last few years, the first detectors have been built to search for some of these hypothesized particles. The experiments are extremely difficult, owing to the shyness of the particles, and it is estimated that detectors of the future need approximately one hundred times more sensitivity before the particles can be found—if they exist.

The Origin of the Universe

The goal of physics is to explain nature with as simple a theory as possible, and the goal of cosmology is to explain the structure and evolution of the universe in terms of that theory. Astronomers and physicists can start with the observed universe and go backwards in time, toward the big bang, using the known laws of gravity and electricity and nuclear forces. The universe contracts. Galaxies turn to aimless blobs of gas. The gas blobs merge. Stars come undone, and eventually so do atoms. The temperature mounts. The nuclei of atoms disintegrate. Then, at about 0.00001 second before the big bang, the subatomic particles filling the universe have energies equal to the highest energies tested in particle accelerators on Earth. Further extrapolation toward the big bang enters the realm of speculation. Until recently, most cosmologists have not worried much about those first few moments. The question of the origin of the universe is almost too big to contemplate.

Today, that question is being contemplated. Astronomers and physicists today believe that many properties of the present universe probably depend on what happened during the first one-hundred-thousandth of a second.

One such property, ironically, is the apparent uniformity of the universe on the large scale, as evidenced by the extreme uniformity of the cosmic background radiation. Although such uniformity and homogeneity are *assumed* in the big bang model, they still must be *explained*, or at least be made plausible.

It seems unlikely to many scientists that the universe would have been created so homogeneously. Initial inhomogeneities might have eventually smoothed themselves out, in the same way that hot and cold water in a bath tub will come to the same temperature by mixing together and exchanging heat. Heat exchange, however, takes time. Since nothing can travel faster than light, the speed of light sets the maximum distance over which heat exchange can take place. Thus, when the universe was 300,000 years old, the maximum distance over which homogenization should have taken place was about 300,000 light-years. Yet the regions of space that produced the cosmic background radiation at that time were about 50 million light-years in size—much too big to have had time since the big bang to exchange heat and homogenize. The problem of accounting for the large-scale uniformity of the universe has been called the horizon problem. It was first clearly stated by Charles Misner of the University of Maryland in the late 1960s. Its solution probably requires an understanding of the first one-hundred-thousandth of a second.

In the late 1970s and early 1980s, a revolution occurred in theoretical cosmology. New ideas from the theory of subatomic particles and forces were brought to bear on an understanding of the birth of the universe. And with these new ideas, new questions could be asked. Before the 1970s, scientists accepted as givens many of the properties of the universe, such as the existence of galaxies and indeed the existence of matter itself, the prevalence of matter over antimatter, and the extreme smoothness of the cosmic background radiation. Before the 1970s, cosmologists were concerned more with measuring the distances and properties of galaxies, measuring the rate of expansion of the universe, and working out the consequences of the big-bang theory. Since the 1970s, armed with new subatomic physics, cosmologists have seriously begun to challenge everything taken before as a given. Why should matter exist at all? Where did the very first quarks and electrons come from? Why was the infant universe apparently so homogeneous? Subatomic physics has pushed back the frontiers of cosmology to much earlier than the first 0.00001 second.

Even before the revolution of the late 1970s, which will be further discussed shortly, cosmology had connected with subatomic physics. Theories of elementary particles and forces depend crucially on how many types of elementary particles there are. One class of subatomic particles is called the lepton. Only three types of leptons are known—electrons, muons, and taus, and their associated antiparticles and neutrinos—but some theories of particle physics predict that there could be many more. Now comes the cosmology. According to theoretical calculations first done in 1964 by British physicists Fred Hoyle and Roger Tayler and, independently, by Soviet physicist V. F. Shvartsman, and then repeated independently and in more detail in 1977 by Gary Steigman of the Bartol Research Foundation, David Schramm of the University of Chicago, and James Gunn of Princeton, the amount of helium produced in the nuclear reactions of the early universe should have depended on the number of types of leptons. The more types of leptons, the more helium. The calculations indicated that for the observed abundance of helium today, about 24 percent of all ordinary matter, there could exist *at most* one new type of lepton beyond the three already known. These were purely theoretical calculations. They were based on the picture of the universe at one minute old as portrayed by the big bang model. Yet the calculations made a prediction about a fundamental property of our universe at the subatomic level. The prediction was tested in 1989. Experiments carried out in the giant electron-positron collider at CERN in Geneva and at the Stanford

Linear Accelerator Center in California indicated that there are *no* new types of leptons. There are only three. This confirmation increases confidence both in our understanding of subatomic physics and in the big-bang theory.

The fruitful union of subatomic physics and cosmology in the 1970s was brought about in large part by the new "grand unified theories" of physics, nicknamed GUTs. These theories propose that there is only a single fundamental force of nature. Under everyday conditions, however, that single force appears as four separate forces: the gravitational force, the electromagnetic force, and two types of nuclear forces. Some of the GUT theories include gravity and some do not. The proposed unity of the four forces is a bit like the unity of the three forms of water. Ice, liquid water, and steam appear very different. Yet they are composed of the same molecules. Furthermore, under suitable conditions, such as a very hot container or a very cold container, the three states of water will combine into a single state.

Since the ancient Greeks, physicists have sought minimalist explanations of nature. Although such theories are extremely appealing, there is still little experimental evidence that any grand unified theory is correct. Any such theory is difficult, if not impossible, to test. The required heat under which the forces would combine—something like 10^{28} degrees—is far higher than the highest temperature that can be created in the laboratory. It is far higher than the hottest temperature at the center of any star. In fact, such a scorching temperature existed only once, in the hot sea of energy that filled the universe when it was much less than a second old. Having no laboratory in the current universe to test their new theories, subatomic physicists were obliged to contemplate the infant universe and thus to journey into the realm of cosmology.

In the early 1980s, subatomic physicists Alan Guth of the Massachusetts Institute of Technology (then at Stanford), Paul Steinhardt and Andreas Albrecht of the University of Pennsylvania, and Andrei Linde of the Lebedev Physical Institute in Moscow—all experts on GUTs—proposed a modification to the big bang model that provided a natural explanation of the horizon problem and other unsolved problems in cosmology. The new cosmological model is called the inflationary universe model, and it has caused a major change in cosmological thinking.

According to the inflationary universe model, when the universe was only about 10^{-35} seconds old—the point in time when the unified force began to condense into its separate forms—the infant universe went through a brief and extremely rapid expansion. The rapid expansion was caused by a strange type of energy that may have existed at this time, an energy causing gravity to repulse rather than attract. By the time the universe was still a tiny fraction, perhaps 10^{-32}, of a second old, the period of rapid expansion, or inflation, was over. The universe then returned to its far slower expansion. The epoch of rapid expansion could have taken a patch of space so tiny that it had already homogenized and quickly stretched it to a size larger than today's entire observable universe. Thus, the inflationary expansion would have homogenized the universe over an extremely vast region, far larger than any region from which we have data. Regions of space that appear to have never been close enough to have exchanged heat, according to extrapolations into the past based on the standard big bang model, were actually much closer, based on the inflationary universe model.

The inflationary universe model makes specific predictions about the formation of structures in the universe. In particular, processes associated with the grand unified force in the early universe would have determined the nature of the initial inhomogeneities, the primeval mounds on the desert that much later condensed into galaxies and groups of galaxies. These predictions have been incorporated into a detailed theory of large-scale structure called the "cold dark matter" model, which is actually a specific case of the gravitational hierarchy models mentioned earlier. The cold dark matter model has been highly influential and has been regarded by most cosmologists as the leading model of cosmic structure formation for the last decade. Many of the computer simulations now used to understand the observed positions and motions of galaxies begin with the initial inhomogeneities of matter decreed by the cold dark matter model.

Unfortunately for theorists, observational evidence against the cold dark matter model has been steadily accumulating and has now

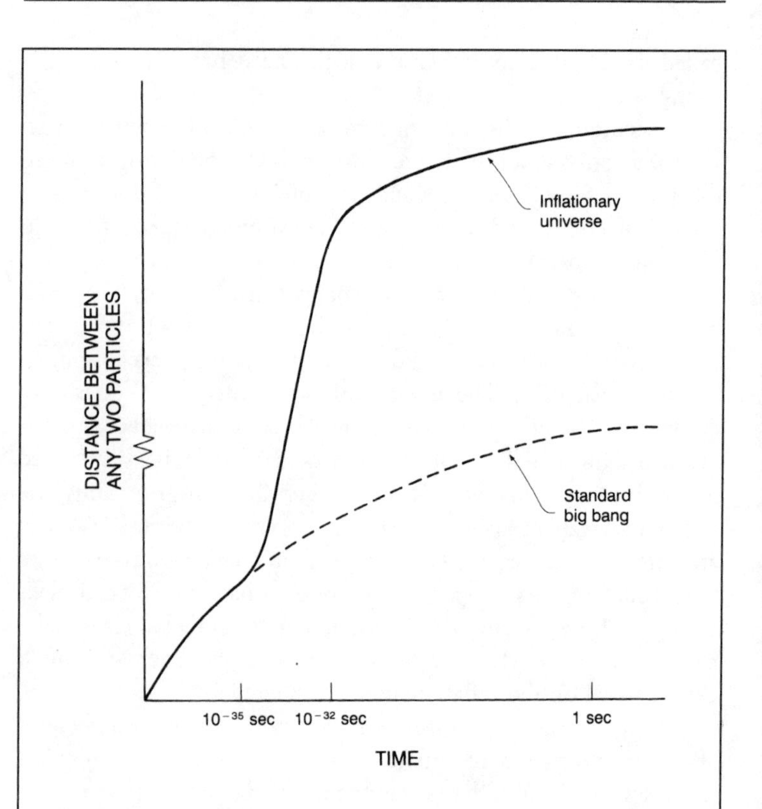

Fig. 18. The expansion of the universe in the standard big bang model and in the inflationary universe model. The expansion may be measured by the distance between any two distant particles. (Galaxies did not exist at these early epochs.)

reached a near-fatal level. The essential problem is that actual galaxies cluster together on large scales much more than can be accommodated by the theory. In the early 1980s, Neta Bahcall of Princeton University and Raymond Soneira of AT&T Bell Laboratories concluded that clusters of galaxies tend to bunch together substantially on scales of several hundred million light-years and larger. In the late 1980s, the discovery of the Great Attractor, mentioned earlier, also indicated more clustering of matter on large scales than can be explained by the cold dark matter model. More recent work by George Efstathiou, S. J. Maddox, and collaborators at Oxford University, and by a collaboration of scientists from Queen Mary and Westfield Colleges, the University of Durham, Oxford University, and the University of Toronto, show that on scales larger than 30 million light-years galaxies cluster together far more than can be easily explained by the cold dark matter model. Many cosmologists now feel that the cold dark matter model is dead. And there don't seem to be any other good candidates to replace it. Badly needed are larger surveys of galaxies, more theoretical ideas, and more complete computer simulations. At the moment, cosmologists are stumped and troubled over the new observations.

It is still conceivable that some version of the inflationary-universe model could be right even if the cold dark matter model is wrong. Whether the inflationary universe model is correct or not, it has given scientists a means to calculate how the universe might have behaved at an extremely early time after the big bang. Scientists now have a role model for how to explain some perplexing properties of the universe—such as the uniformity of the cosmic background radiation—in terms of calculable physical processes.

Suppose for the sake of argument that the inflationary universe model is right. That takes us back to 10^{-35} seconds after the big bang. But what happened before then? Some physicists believe that many of the properties of our universe were in fact determined in the first 10^{-43} seconds, in the so-called Planck era. In the Planck era—named after Max Planck, one of the founders of the quantum theory—the entire universe would have been subject to the same "quantum mechanical fluctuations" as subatomic particles. Quantum mechanics, developed in the 1920s and 1930s and later confirmed by experiment, describes the behavior of matter at the subatomic scale. That behavior is extremely counterintuitive to us macroscopic creatures. According to quantum mechanics, nature at the subatomic scale has an intrinsic fuzziness; nature can be described only by probabilities, not certainties. For example, an electron behaves as if it were occupying several places at the same time.

In the Planck era, large quantities of matter and energy could have done the same, appearing and disappearing wholesale. The concept of time itself may have lacked any meaning. For all practical purposes, the Planck era may be considered the origin of the universe.

In the last decade, theoretical physicists, led by Stephen Hawking of Cambridge University, have attempted to calculate the expected behavior of the universe in the Planck era. Such work is called quantum cosmology. Hawking begins with the concepts of quantum mechanics and general relativity theory, makes some broad assumptions about the shape of the universe in a fictitious, higher-dimensional space, and then works out the consequences. Hawking is trying to calculate the birth of the universe. Admittedly, his calculation is oversimplified. But it may have some of the right ingredients. If such a calculation could ever be done reliably, scientists would not be forced to assume anything about the initial conditions of the universe. We would know why the universe is what it is. Quantum cosmology is our Anu and Nudimmut.

The End of the Universe

We know that the universe is expanding now, but will it keep expanding forever? What is its ultimate fate? Almost certainly, the expansion is slowing. Just as a rock thrown upward from the Earth continuously slows as it rises, retarded by the gravitational pull of the Earth, the universe should be slowing down in its outward expansion by virtue of its own attractive gravity. The competition between the outward motion of expansion and the inward pull of gravity leads to two possibilities: the universe could expand forever, like a rock thrown upward with sufficient speed to escape the gravity of the Earth, or the universe could expand to a point and then begin collapsing, like a rock that reaches a maximum height and then starts falling back to Earth. The two possibilities are called open universes and closed universes, respectively. Open universes last forever. Closed universes have an end in time, when the universe has collapsed back to a condition of near-infinite density, in a kind of reverse big bang.

Which path our universe will take depends on how the cosmic expansion began, in a similar way that the rock's path depends on its initial speed relative to the strength of the Earth's gravity. Even without knowledge of these initial conditions, however, we can determine whether our universe is open or closed by comparing its *current* rate of expansion to its *current* average density of matter. If the density is greater than a critical value, then gravity dominates

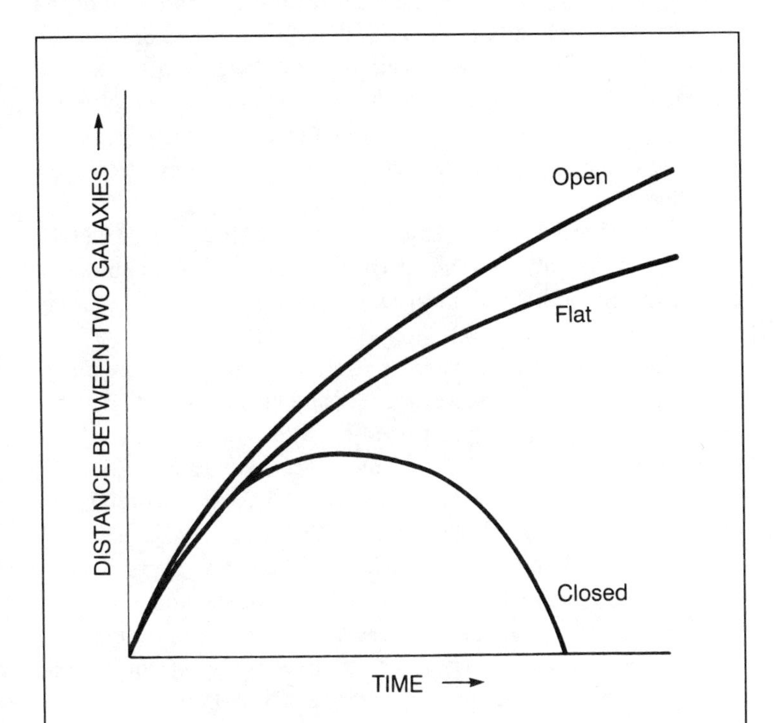

***Fig. 19.** Expansion of the universe in time for closed, open, and flat cosmologies. The expansion may be measured by the distance between any two distant galaxies. In a closed universe, the universe expands at first and then contracts.*

and the universe is closed, fated to collapse at some time in the future. If the density is less than the critical value, the universe is open. The critical value of density, in turn, is determined by the current rate of expansion of the universe. That current rate has been measured to be a doubling of size every 10 billion years, which translates to a critical density of about 10^{-29} grams per cubic centimeter, the density achieved by spreading the mass of a personal computer through a volume the size of the sun. The ratio of the actual density of matter to the critical density is called omega. Restating the possible fates of the universe in terms of omega, we can say the universe is open if omega is smaller than one and closed if omega is larger than one. In the special case that omega is exactly one, the universe is said to be flat—precisely midway between open and closed.

According to various measurements, the actual average density of matter in the universe is about 10^{-30} grams per cubic centimeter, or about one-tenth the critical value. Omega is about 0.1. Our universe seems to be open.

Unfortunately, the matter isn't closed. Omega is difficult to measure. If the universe were precisely homogeneous, so that all its pieces had exactly the same properties, the value of omega could be determined rather easily, from measurements nearby. Take any nearby volume of space where distances can be well determined, figure out the mass contained within by its gravitational effects, and divide by the volume to get the density. Then, use the redshift to measure the recessional speed of the edge of the volume, divide by the distance, and get the rate of cosmic expansion and the corresponding critical density. However, the universe is not homogeneous. Local inhomogeneities cause the density of the universe and the rate of expansion to vary from one place to another. The best we can do is to hope that these inhomogeneities disappear on a large enough scale and make our measurements on such large scales.

One possible means of making such measurements involves the scattering of the cosmic background radiation by hot gas in clusters of galaxies, an effect first pointed out by Yakov B. Zel'dovich and Rashid Sunyaev of the Soviet Union. The hot gas gives a slight energy boost to the radio waves as they pass through the gas on their way to Earth. By measuring both the change in energy of the radio waves and the X-ray emission from the hot gas, the distance to the cluster of galaxies may be well determined. Such measurements repeated for a large number of galaxy clusters would permit a more accurate determination of the rate of expansion of the universe on large scales. With the Millimeter Array, the Advanced X-ray Astrophysics Facility, and other instruments, astronomers hope to make these measurements in the coming decade.

Likewise, studies of the velocities and distances to a large number of galaxies could pin down the local values of both omega and the Hubble constant. Peculiar velocities of galaxies depend upon the extra amount of matter concentrated in a region, over and above the average density of cosmic matter. Measurement of peculiar velocities, together with a knowledge of how much matter there is above the average, leads to an estimate for omega.

Omega is not completely uncertain. Astronomers have identified enough matter in moderately large volumes of space so that omega cannot be less than 0.1. On the other end, if omega were larger than 2, the estimated age of the universe would be less than the age of the Earth. Thus omega almost certainly must lie between 0.1 and 2. Cosmologists would be happier if they knew that omega was definitely smaller than 1 or definitely larger than 1, answering the question of whether the universe is closed or open. Perhaps we will know in the next decade.

The inflationary universe model firmly predicts that omega should be equal to 1, exactly. On this basis, the model can in principle be ruled out or supported from observational evidence. At the present time, the evidence points to a value of omega closer to 0.1. Thus, scientists who believe on theoretical grounds that the inflationary universe model is right must have faith that an enormous amount of mass is hiding from us. Such missing mass is called "missing mass."

To summarize, the light-emitting matter we see accounts for sufficient mass to make omega about 0.01; the unseen but gravitationally detected dark matter accounts for another factor of 10 of mass, increasing omega to about 0.1. Advocates of the inflationary universe model, which requires an omega equal to 1, must hypothesize that every cubic light-year of space contains on average yet ten times more mass, not only unseen but so far undetected.

Although the inflationary universe model is at odds with current observations, its general features and explanatory powers have captured the minds of many scientists. It is sobering to realize that the highly influential inflationary universe model was unknown little more than a decade ago. Like Supernova 1987A, the idea exploded. We should expect similar explosions in the future.

If the universe is closed, it will one day stop expanding and begin to collapse, in a kind of reverse big bang. Temperatures will begin mounting, instead of dropping, and all matter will eventually disintegrate and get compressed into oblivion. Whether a new universe will arise after the old one's demise, as in each dawn of the life of Brahma, is completely unknown.

Alternatively, if the universe is open or flat, it will keep expanding for eternity, growing colder and colder and ever more diffuse. Stars and galaxies will continue to evolve, but more and more slowly. According to estimates, after about 10^{14} (100 trillion) years all the stars will have burned out and dimmed, after about 10^{15} years planets will have come loose from their central stars, after about 10^{19} years stars will have come loose from their galaxies, and after about 10^{1500} years all matter will have decayed into iron. There will be plenty of time for all of these things to occur. Some scientists, notably Freeman Dyson of the Institute for Advanced Study in Princeton, believe that life could continue in such a diminishing world. As the vast reaches of space grow empty and cold, the slowing rate of messages between neurons may be compensated for by the enormous time available, stretching and unwinding into infinity. Years will become seconds.

Finally, we must not forget that all we can do with certainty is make maps of our local region of space. Even if the cosmos is infinite in extent, only a limited volume is visible to us at any moment: we can see only as far as light can have traveled since the big bang. As we look farther into space, we see light that has traveled longer to reach us. Eventually, at some distance, the light just now reaching our telescopes was emitted at the moment of the big bang. That distance marks the edge of the currently observable universe, some 10 to 20 billion light-years away. We cannot see farther because there hasn't been time for light to have traveled from there to here. And we have no way of knowing what lies beyond that edge. It is conceivable that extremely distant regions of the cosmos could have different forces, different types of particles, even different dimensionalities of space. If so, it would be impossible for us ever to witness more than a tiny portion of the tapestry of nature.

By the same token, there is surely much to surprise us. As the reality of an expanding universe was completely unknown in 1920 and the reality of quasars unknown in 1960, who can imagine what astronomers will find by the year 2000? In the final words of *The Origin of Species*, Darwin wrote "from so simple a beginning, endless forms most beautiful and most wonderful have been, and are being evolved." How much more true for the universe.

Some Recent and Proposed Astronomical Instruments

NAME	DATES OF OPERATION*	WAVE-LENGTH	LOCATION	COMMENTS
Arecibo	1960–	radio	Puerto Rico	world's largest
Kitt Peak 4-meter	1973–	visible and infrared	Arizona	national access
MMT	1978–	visible	Arizona	6 segments
Cerro Tololo 4-meter	1978–	visible and infrared	Chile	
IUE	1978–	ultraviolet	space	
Einstein	1978–1981	X-ray	space	
VLA	1980–	radio	New Mexico	
IRAS	1983	infrared	space	all-sky survey
Ginga	1987–1992	X-ray	space	Japanese

NAME	DATES OF OPERATION*	WAVE-LENGTH	LOCATION	COMMENTS
COBE	1989–	radio infrared	space	
HST	1990–	visible ultraviolet	space	to have infrared capability
ROSAT	1990–1995	X-ray	space	joint German, U.S., U.K.
GRO	1991–2005	gamma ray	space	
VLBA	1992–	radio	continental U.S.	
Keck 10-meter	1992–	visible infrared	Hawaii	
Smithsonian 6.5-meter	1993–	visible infrared	Arizona	upgrade of MMT
GONG	1993–	visible	global	solar monitor
GBT	1995–	radio	West Virginia	
Submillimeter Array	1996–	radio infrared	Hawaii	Smithsonian
SOFIA	1997–2017	infrared	airborne	
Columbus 8-meter	1996–	visible infrared	Arizona	two 8.4-meters
Magellan 8-meter	1996–	visible	Chile	
Japanese 7.5-meter	1996–	visible	Hawaii	
IRO	1998–	infrared visible	Hawaii	national access
VLT	1998–	visible infrared	Chile	four 8-meters European
AXAF	1998–2013	X-ray	space	
SIRTF	2000–2005	infrared	space	
MMA	2001–	radio	New Mexico	

NAME	DATES OF OPERATION*	WAVE-LENGTH	LOCATION	COMMENTS
FUSE	1997–?	ultraviolet	space	
EUVE	1992–?	ultraviolet	space	
OSL	1998–?	X-ray visible	space	
LEST	1998–?	visible infrared	Canary Islands	
LST	2010–2020	visible	space	

*All dates in the future are uncertain.

ACRONYMS USED

AXAF: Advanced X-ray Astrophysics Facility
COBE: Cosmic Background Explorer
EUVE: Extreme Ultraviolet Explorer
FUSE: Far Ultraviolet Spectroscopy Explorer
GBT: Green Bank Telescope
GONG: Global Oscillations Network Group
GRO: Gamma Ray Observatory
HST: Hubble Space Telescope
IRAS: Infrared Astronomical Satellite
IRO: Infrared Optimized Telescope
IUE: International Ultraviolet Explorer
LEST: Large Earth-Based Solar Telescope
LST: Large Space Telescope
MMA: Millimeter Array
MMT: Multiple Mirror Telescope
OSL: Orbiting Solar Laboratory
ROSAT: Roentgen Satellite
SIRTF: Space Infrared Telescope Facility
SOFIA: Stratospheric Observatory for Far-Infrared Astronomy
VLA: Very Large Array
VLBA: Very Long Baseline Array
VLT: Very Large Telescope

Illustration Credits

Cover Illustration. Courtesy of National Optical Astronomy Observatories.

Fig. 1. Drawing by Alan Lightman.

Fig. 2. Courtesy of JPL/NASA.

Fig. 3. Courtesy of JPL/NASA.

Fig. 4. Courtesy of B. A. Smith and R. J. Terrile, *Science*, Vol. 226, p. 1421 (1984).

Fig. 5. Courtesy of NASA.

Fig. 6. Big Bear Solar Observatory, California Institute of Technology.

Fig. 7. National Radio Astronomy Observatory.

Fig. 8. Drawing by Alan Lightman.

Fig. 9. Drawing by Alan Lightman.

Fig. 10. Courtesy of TRW and NASA.

Fig. 11a. Courtesy of Rudoph Shield and the Smithsonian Astrophysical Observatory.

Fig. 11b. Courtesy of Rudolph Shield and the Smithsonian Astrophysical Observatory.

Fig. 12. Courtesy of Harvard College Observatory.

Fig. 13. Photo by Fritz Zwicky, courtesy of Halton Arp.

Fig. 14. Courtesy of the Museum of the History of Science, Florence.

Fig. 15a. Courtesy of J. A. Biretta and F. N. Owen.

Fig. 15b. Courtesy of J. A. Biretta and F. N. Owen.

Fig. 16. Courtesy of M. J. Geller and J. P. Huchra, *Science*, Vol. 246, p. 897 (1989).

Fig. 17. Reprinted by permission of the publishers from *Origins: The Lives and Worlds of Modern Cosmologists*, by Alan Lightman and Roberta Brawer (Cambridge, Mass.: Harvard University Press), copyright © 1990 by Alan Lightman and Roberta Brawer.

Fig. 18. See Fig. 17.

Fig. 19. See Levitt photo.

Elliot photo by Alan Lightman.

Baliunas photo by Steve Padilla.

Levitt photo. Reprinted by permission of the publishers from *Ancient Light*, by Alan Lightman (Cambridge, Mass.: Harvard University Press), copyright © 1991 by Alan Lightman. *Ancient Light* is adapted from *Origins: The Lives and Worlds of Modern Cosmologists*, by Alan Lightman and Roberta Brawer (Harvard University Press, 1990).

Illingworth photo courtesy of Garth Illingworth.

Peebles photo by Robert P. Matthews, Princeton University.

Index

Page numbers in *italics* refer to illustrations.

Albrecht, Andreas, 98–99
active galaxies:
  gaseous jets of, 67–69, *69*
  magnetic fields of, 67–68
  power source of, 62–69
adaptive optics, 19–20
"Adonais" (Shelley), 32
Advanced X-ray Astrophysics Facility (AXAF), x, 44, *45*, 46, 47, 66, 67, 94, 105
Alpha Centauri, 15, 16
Alpher, Ralph, 82
American Astronomical Society, 91
Andromeda galaxy, 54, 58, 66, 91
  orbital speed of hydrogen in, *90*
angular resolution, 16, 17, 18, 20–21, 26, 32, 66, 67, 68
  defined, 20
antiparticles, 66
Anu, xv
Arecibo Telescope, 72, 86
Aristotle, 1, 3, 77, 79
Astro D, 94
astrolabe, xvi
astronomers, modern vs. past, xv
astronomical instruments:
  judgment criteria for, 20–21
  recent and proposed, 109–111
  see also specific instruments
astronomy:
  as first science, xv–xvi
  technological change and, xvi–xviii
  theoretical, xvii, 2, 11–12, 36, 38
  see also specific topics

Astronomy and Astrophysics Survey Committee, ix–x
atmosphere, 6, 10
  of Earth, 6, 7, 10, 16, 19–20, 44
  of Venus, 6–7, 8
atoms, xv, 1, 36, 43, 81, 82
Australia, 57
axions, 95

Baade, Walter, 36
Babylonian creation story, xv
Bahcall, John, ix–x, 27
Bahcall, Neta, 100–1
Baksan Laboratory, 29
Baliunas, Sallie, 39–40, 39
barred spiral galaxy NGC 3992, *50*
Bell, Jocelyn, 36
Bell Laboratories, 72
Beta Pictoris, *14*
big bang model, 77–83, 95–98, *100*
  evidence in support of, 81
  subatomic physics and, 96–98
blackbody radiation, 82–83
black holes, 41, 42–43, 63–66
  concept of, 63–64
Bosma, Albert, 55
Broadhurst, T. J., 85
Bruno, Giordano, 13
Burstein, David, 86

California, University of (Berkeley), 73
California, University of (Santa Barbara), 73
California Institute of Technology, 73
Canada, 29
Cannon, Annie Jump, 53–54
Cape Verde Islands, 3
carbon, xvi, 4, 21, 34, 82
carbon dioxide, 6
carbon monoxide, 4, 56
Carnegie Institution, 55, 91
Central Bureau for Astronomical Telegrams, 42
Cepheid variables, 51, 52, 54, 80, 87
CERN, 97–98
chaos theory (nonlinear dynamics), 11
charge-coupled devices, xvii
Charon, 8
chemical energy, 24
Chicago, University of, 85–86
"Christabel" (Coleridge), 32
Christy, James, 8
climate, of Earth, 6, 10, 25–26
clocks, xvi
closed universe, 102–4, *103*, 106
cobalt, 43
cold dark matter model, 99–101
Coleridge, Samuel Taylor, 32, 34
Comet Rendezvous Asteroid Flyby, 4
comets, xvi, 4
computers, xvii, 2, 12, 42, 43, 69
  chaos theory and, 11
  cold dark matter model and, 99
  evolution of galaxies and, 59
  guide stars and, 20
  large-scale structures and, 88–89
  miniaturized, 45
  in search for extraterrestrial intelligence, 22
Congress, U.S., x
continental drift, 3
Copernicus, Nicholas, xvi, 78, 79
Cornell University, 8
Cosmic Background Explorer, 72–73, 83

cosmic background radiation, 72–73, 89, 93
  big bang and, 81, 83
  discovery of, 72, 83
  measuring omega and, 104–5
  uniformity of, 86, 96, 101
cosmology, 77–107
  of ancients, xv, 77
  cold dark matter model, 99–101
  inflationary universe model, 99, *100*, 101, 105–6
  Judeo-Christian, 77–78
  large-scale structures and, 83–89
  quantum, 102
  see also big bang model
Cowie, Lennox, 71
Cygnus X-1, 65

dark matter, 86–87, 89–95
  as haloes of galaxies, 55
  mapping of, 94
Darwin, Charles, xvi, 3, 107
Davies, Roger, 86
Davis, Raymond, 28
Democritus, 1
De Revolutionibus (*A Perfit Description of the Caelestiall Orbes*) (Copernicus), 78
de Sitter, Wilhelm, 79
deuterium, 82
Dicke, Robert, 82, 92–93
Digges, Thomas, 78
distances, cosmic, methods for measurement of, 87–88
Doppler shift, 16–17, 27, 60, 79–80, 84–87, *84*
Doroshkevich, A. G., 71
Drake, Frank, 21
Dressler, Alan, 86
Dürer, Albrecht, xvi
Durham, University of, 101
Dyson, Freeman, 106

Earth, 1–3
  age of, 3, 81
  atmosphere of, 6, 7, 10, 16, 19–20, 44
  climate of, 6, 10, 25–26
  gravity of, 64, 103
  mass of, 2
  other planets compared with, 6–8, 15
earthquakes, solar, 26
Eddington, Arthur, 24, 36
Efstathiou, George, 101
Einasto, J., 55
Einstein, Albert, 63, 78–79, 80
Einstein X-ray Observatory, xvii–xviii, 44, 67
electromagnetic spectrum, wavelengths of, 4–6, *5*, 10, 15, 16–17
electronic light detectors, xvii
electrons, 36, 40, 66, 81, 101
Elliot, James, 8, 9, *9*
Ellis, R. S., 85
energy:
  solar, 3, 23–24
  of quasars, 60–63
Enuma Elish, The, xv
escape speed, 64
European Space Agency, 18
evolution, xv, xvi
  of galaxies, xvi, 56–62
  of stars, 32–47, 58
  of sun, 35
extraterrestrial intelligence, search for, 21–22
Extreme Ultraviolet Explorer, 47

Faber, Sandra, 86, 87
Far Ultraviolet Spectrosocopy Explorer, 47
Fisher, J. Richard, 87
flat universe, *103*, 104, 106
Fleming, Williamina, 53–54
Ford, W. K., *90*
fractional variation method, 88
Friedmann, Alexander, 79, 80

galaxies:
  active, 62–69, *69*
  Andromeda, 54, 58, 66, *90*, 91
  birth of, 61–62, 70–73
  clusters of, xvi, 58, 100–1
  collisions of, 61
  dark matter halo of, 55
  discovery of, 49–56, 62
  evolution of, xvi, 56–62
  gravitational hierarchy model of, 70, 71, 72, 93, 99
  interactions of, 57, 58, 61
  intergalactic medium and, 73
  location of orbiting hydrogen in, 55–56
  pancake model of, 71–72
  radio, 56, 62
  redshifts of, 79–80, 84–87, *84*
  spiral, *50*, 55; *see also* Milky Way
  stars in, 49–51, 52 54, 58, 60, 62
  surveys of, 85–86
  velocities of, age of universe and, 80–81
  "wall" of, 84, *84*, 85
Galileo, 1, 25, 49, 59
Galileo mission, 8
Gamma Ray Observatory (GRO), x, 43, 46, 66
gamma rays, 5, *5*, 6, 34, 66
  supernovae and, 43–44, 46
  wavelengths of, 43
Gamow, George, 82
Geller, Margaret, 83–84, *84*
general relativity theory, 78–79
geology, 3, 10
Germany, 94
Global Oscillation Network Group (GONG), 27, 40
Goddard Space Flight Center, 73
Goldreich, Peter, 12
grand unified theories (GUTs), 98
Gran Sasso Laboratory, 29
gravitational hierarchy model, 70, 71, 72, 93, 99
"gravitational lens" phenomenon, 94
gravity, xvi, 15, 16
  in birth of galaxies, 70, 71, 72
  of black hole, 41, 63, 66
  dark matter and, 90–91
  of Earth, 64, 103
  Einstein's theory of, 63, 78–79
  end of universe and, 102–4
  escape speed and, 64
  of galaxies, 57, 58
  in GUTs, 98
  in inflationary universe model, 99
  Newton's theory of, xvi, 12, 63, 64, 78
  in star formation, 29
  in stellar evolution, 34, 35, 38, 40, 41
Great Attractor, 86, 101
Great Britain, 29, 57
Great Red Spot, 10–11
"Great Wall," 84, *84*, 85
Green Bank Telescope (GBT), 31, 72, 86
greenhouse effect, 6
Gunn, James, 86, 97
Guth, Alan, 98–99

Hale telescope, 17, 19
Halley's comet, 4
Harvard College Observatory, 53, 54
Harvard-Smithsonian Center for Astrophysics, 39
Hawking, Stephen, 102
Heinlein, Robert, ix
helioseismology, 26–27, *28*, 40
helium, 2, 34, 46, 93, 97
  big bang and, 81, 82
Herman, Robert, 82
Herschel, William, 49
Hertzsprung, Ejnar, 32, 34, 36, 38
Hertzsprung-Russell diagram, 32–34, *33*
Hewish, Anthony, 36
Hipparchus, 49
Horowitz, Paul, 22
Hoyle, Fred, 82, 97
Hubble, Edwin, xvi, 54, 79–80
Hubble constant, 80, 85, 105
Hubble Space Telescope (HST), x, xi, 16, 35, 60, 61, 62, 66
Huchra, John, 83–84, *84*
hydrogen, 2, 4, 29, 34, 35, 46
  big bang and, 81, 82
  orbital speed of, *90*
  orbiting, location of, 55–56

Illingworth, Garth, 74–75, *74*
inflationary universe model, 99, *100*, 101
  omega and, 105–6
infrared astronomers, xvii
Infrared Astronomical Satellite (IRAS), 3, 17, 21, 30, 61
Infrared Optimized telescope (IRO), 18–20, 31, 62
infrared radiation, xvii, *14*, 15, 34
  of active galaxies, 62
  of quasars, 61
  star formation and, 29, 31, 62
  wavelength of, 4, 5, *5*, 17
Infrared Satellite Observatory, 18
infrared-sensitive array, 19
infrared telescopes, 60, 61
  IRAS, 3, 17, 21, 30, 61
  IRO, 18–20, 31, 62
  SIRTF, x, 4, *14*, 15, 17–18, *18*, 31, 35, 47, 61, 62, 95
  SOFIA, 10, *14*, 31, 46–47
Institute for Advanced Study, 85–86
instruments, *see* astronomical instruments; *specific instruments*
"interferometric" space telescopes, 16, 68, 87
intergalactic medium, 73
International Ultraviolet Explorer, 42, 46
Io, 8
iron, 35, 43, 82

Jackson, Robert E., 87
Janksy, Karl, 56
Japan, 68, 94
Jeans, James, 70
Jones, Albert, 42
Jovian planets, 2
Judeo-Christian cosmology, 77–78
Jupiter, 1, 2, 6
  gravity of, 16
  Great Red Spot of, 10–11
  other planets compared with, 15
  volcanoes on moon of, 6, 8

Kaasik, A., 55
Kant, Immanuel, 2
Keck telescope, 20
Kepler, Johannes, 42
Kirshner, Robert, 42

Koo, David C., 85
Kron, Richard, 85

Laplace, Pierre-Simon, 12, 64
Large Earth-Based Solar Telescope (LEST), 26
Large Magellanic Cloud, 42
Large Space Telescope (LST), 60
lead, 3
Leavitt, Henrietta Swan, 51, 53–54, 53
leptons, 97–98
light:
  from galaxies, 58–59
  speed of, 58
  see also visible light
light-years, 15, 54
Linde, Andrei, 98–99
lithium, 82
luminosity-stellar speed relationship, 87–88
Lyell, Charles, 3
Lynden-Bell, Donald, 86

Maddox, S. J., 101
Magellan mission, 8
magnetic fields:
  of active galaxies, 67–68
  of neutron stars, 40
  of sun, 25, 26, 40
"main sequence" of stars, 32–35, 33, 37
Marcus, Phillip, 11
Mars, 1, 2, 6, 10
  search for life on, 7–8
Marsden, Brian, 42
mass:
  of Great Attractor, 86, 101
  missing, 105
  peculiar velocities and, 86–87, 105
  of planets, 1–2, 8
  solar, 1, 2
  of stars, 1, 2, 30, 33, 36, 41, 42–43
  of universe, 70
Massachusetts Institute of Technology, 73
matter, 29
Maunder, Edward, 26
Maunder, minimum, 26
M87 (Virgo A), 68, 69
Mercury, 1, 2
methane, 4, 8
Michell, John, 64
microchips, 45
Milky Way, 49–51, 54, 57, 80
Millimeter Array (MMA), 4, 17, 31–32, 31, 56, 105
Misner, Charles, 96
missing mass, 105
molecules, complex, 4
moons, 2, 4
  of Jupiter, 1, 6, 8
  of Pluto, 8
  of Uranus, 6, 12
Mount Wilson Observatory, 54
muons, 97

National Academy of Sciences, ix
National Aeronautics and Space Administration (NASA), 42, 44–45
National Radio Astronomy Observatory, 21
nebular hypothesis, 2–3
Neptune, 1, 2
  rings of, 12
Netherlands, 57
neutrinos, 27–29, 41, 43, 97
neutrons, 36, 81, 82
neutron stars, 35–36, 37, 38, 40
  black holes and, 64–65
  formation of, 40–41

magnetic fields of, 40
Newton, Isaac, 23, 34, 79
gravity theory of, xvi, 12, 63, 64, 78
NGC 3992, *50*
NGC 4565, *50*
NGC 5426, *57*
NGC 5427, *57*
nickel, 41, 43
nitrogen, 4, 7
nonlinear dynamics (chaos theory), 11
novae, 41
Nuclear Astrophysics Explorer (NAE), 43–44, 46
nuclear energy, 24
nucleus, atomic, 36
Nudimmut, xv

Oberth, Hermann Julius, 44
observation, astronomical, role of, 36, 38
omega, 104–5
On the Heavens (Aristotle), 77
open universe, 102–4, *103*, 106
Oppenheimer, Robert, 64–65
optical interferometer telescopes, 87
optics, adaptive, 19–20
Orbiting Solar Laboratory (OSL), 26
Origin of Species (Darwin), 107
Ostriker, Jeremiah, 55, 71
Oxford University, 101
oxygen, xvi, 4, 7, 41, 82

pancake model, 71–72
peculiar velocities, 86–87, 105
Peebles, James, 55, 70, 82, 92–93, *92*
"pencil beam" surveys, 85
Pennsylvania, University of, 28
Penzias, Arno, 72
photinos, 95
photographs, xvii, 53–54
Physical Cosmology (Peebles), 93
physics:
GUTs of, 98
subatomic, 96–99, 101–2; *see also specific subatomic particles*
Pickering, Edward, 53, 54
Planck, Max, 101
Planck era, 101–2
planets, 1–4, 6–21
dark matter as, 95
mass of, 1–2, 8
nebular hypothesis of origin of, 2–3
orbits of, xvi, 12
search for, 13–21, 95
terrestrial vs. Jovian, 2
see also specific planets
Pluto, 1, 2, 8, 9, 10
moon of, 8
positrons, 66
Princeton University, 73, 85–86
Principia (Newton), 23
projects, scientific, size of, x
protons, 36, 81, 82
protoplanetary disks, 2, 3, 17, 19, 31

quantum cosmology, 102
quantum mechanics, 101–2
quasars, xvii–xviii, 73
discovery of, 60, 61
power source of, 62–69
Queen Mary and Westfield Colleges, 101

radiation:
blackbody, 82–83
cosmic background, *see* cosmic

radiation (*cont.*)
background radiation
infrared, *see* infrared radiation
solar, 24, 25
of stars, 24, 25, 34, 38, 47
ultraviolet, xvii, 5, *5*, 25, 26, 34, 47
wavelengths of, 4–6, *5*
radio galaxies, 56, 62
radio telescopes, xvii, 21, 30, 40
Arecibo, 72, 86
evolution of galaxies and, 56–57, 61
Green Bank, 31, 72, 86
radio waves, xvii, 6, 21
of active galaxies, 62
of carbon monoxide, 56
of orbiting hydrogen, 56
of quasars, 61
star formation and, 31, *31*, 32
of stars, 34, 36
wavelength of, 4–5, *5*, 17
Reber, Grote, 56–57
red giants, 34–35, 37, 46
redshifts, 16–17, 60, 79–80, 84–87, *84*
Roberts, Morton, 90
Rubin, Vera, 55, *90*, 91–92
Russell, Henry Norris, 32, 34, 38

Saar, E. 55
Salpeter, Edwin, 65
Saturn, 1, 2, 6
rings of, 6, 12, *13*
Schmidt, Maarten, 60
Schramm, David, 82, 97
Schwabe, Heinrich, 25
Schwarzschild, Karl, 64
Schwarzschild radius, 64
science, "big vs. little," x
scientific notation, 6
Shakespeare, William, 34
Shapley, Harlow, 54
Shelley, Percy Bysshe, 32, 34
Shelton, Ian, 42
"shepherds" theory, 12
Shvartsman, V. F., 97
"sight and grid," xvi
silicon, 41
Sirius, 36
Slipher, Vesto, 80
Snyder, Hartland, 64–65
solar corona, 24–25, 26
solar flare, 25, 26
solar system:
formation and evolution of, 1–12
sun-centered, xvi
see also planets; sun; *specific planets*
solar wind, 10, 25, 26
Soneira, Raymond, 100–1
Soviet-American Gallium Experiment, 29
Soviet Union, 19, 68
Space Infrared Telescope Facility (SIRTF), x, 4, *14*, 15, 17–18, *18*, 31, 35, 47, 61, 62, 95
spectral resolution, 17, 20–21, 44
spiral galaxies, 55
NGC 3992, *50*
NGC 4565, *50*
see also Milky Way
Sporer, Gustav, 26
Stanford Linear Accelerator Center, 97–98
stars, 3, 23–47
central, 13, 15, 16–17; *see also* sun
central, wobbling of, 15–16
Cepheid, 51, 52, 54, 80, 87
defined, 1
evolution of, 32–47, 58
explosion of, xvi, 41–44, 46

formation of, xvi, 29–32, 62
in galaxies, 49–51, 52, 54, 58, 60, 62
guide, 20
luminosities of, 10, 32–34, *33*, 51, 52
"main sequence" of, 32–35, *33*, 37
masses of, 1, 2, 30, *33*, 36, 41, 42–43
measurement of distance to, 51, 54
misfits, 34
neutron, *see* neutron stars
red giants, 34–35, 37, 46
white dwarfs, 35, 36, 37, 40
Steigman, Gary, 97
Steinhardt, Paul, 98–99
Stratospheric Observatory for Far-Infrared Astronomy (SOFIA), 10, *14*, 31, 46–47
structures:
defined, 83
large-scale, 83–89
subatomic physics:
cosmology and, 96–99, 101–2
see also specific particles
Submillimeter Wavelength Telescope Array, 31, 32
Sudbury Neutrino Observatory, 29
sulfur, 4
sun, 2, 4, 13, 23–30, 46, 54
energy source of, 3, 23–24
evolution of, 35
interior of, 26–29, *28*, 40
Jupiter's gravity and, 16
lower luminosity of, 10
magnetic fields of, 25, 26, 40
mass of, 1, 2
sunspots, 25–26
Sunyaev, Rashid, 104
supernovae, 41–44, 46–47
Supernova 1987A, 41–42, 43, 46
Szalay, Alex S., 85

taus, 97
Tayler, Roger, 82, 97
technological change, xvi–xviii
telescopes, x, 1, 3, 8, 16–21, 44–46
angular resolution of, 16, 17, 18, 20–21, 26, 32, 66, 67, 68
evolution of galaxies and, 58–62, *59*
of Galileo, 59
guidance of, xvii
infrared, *see* infrared telescopes
interferometric, 16, 68, 87
large, 19, 20
radio, *see* radio telescopes
see also specific telescopes
Terlevich, R. J., 86
terrestrial planets, 2
Thomson, William, 3
Tinsley, Beatrice, 58
titanium, 43
Tonry, John, 88
Toronto, University of, 101
Tremaine, Scott, 12
Tully, R. Brent, 87
turbulent fluids, 11
Tyson, Anthony, 94

ultraviolet radiation, xvii, 5, *5*, 26
of stars, 25, 34, 47
Uniformitarian school, 3
universe, 77–107
age of, 81
dark matter in, 55, 86–87, 89–95
end of, 102–7
expansion of, xvi, 79–80, 82, 83, 99, *100*, 102–7, *103*

universe (*cont.*)
flat, *103*, 104, 106
homogeneity of, 80, 83–85, *84*, 96, 104
large-scale structure of, 83–89
lumps of, 83
mass of, 70
nonstatic, 79, 80
open vs. closed, 102–4, *103*, 106
origin of, xvi, 95–102
static, 78, 79, 80
see also cosmology
uranium, 3, 81
Uranus, 1, 2, 6
rings of, 8, 9, 12

Venus, xvi, 1
atmosphere of, 6–7, 8
craters on, 7, 8
volcanoes of, 7, 8
Very Large Array, 72
Very Large Telescope, 20
Very Long Baseline Array, 61, 66, 67–68
Very Long Baseline Interferometers (OVLBI), 68
Viking mission, 6, 7–8
Virgo A (M87), *68*, *69*
Virgo cluster of galaxies, 58
visible light, 4, 5, *5*, 6
of active galaxies, 62
visible-light telescopes, 16, 20, 26, 60, 61, 66
galaxy survey with, 86
Hale, 17, 19
Hubble Space, x, xi, 16, 35, 60, 61, 62, 66
interferometric, 16
largest, 19
volcanoes, 6, 7, 8
Voyager spacecraft, 6, 8, 10–11, 12, *13*
V-2 rockets, 44

Wagoner, Robert, 82
wavelengths:
of electromagnetic spectrum, 4–6, *5*, 10, 15, 16–17
of gamma-ray emissions, 43
of sound, 16, 21–22
Wegener, Alfred, 3
Wegner, Gary, 86
white dwarfs, 35, 36, 37
formation of, 40
Wilson, Robert, 72
Wisdom, Jack, 12

X-ray astronomers, xvii–xviii
X-ray background, 67
X-ray Roentgen Satellite, 94
X-rays, xvii, 5, *5*, 25, 26, 34, 44, 47
of active galaxies, 62, 67
of quasars, 67
supernovae and, 43, 46

Yahil, Amos, 55

Zel'dovich, Yakov B., 65, 70–71, 82, 104
Zwicky, Fritz, 36, 90–91